GRANVILLE SHARP PATTISON

History of American Science and Technology Series

General Editor, LESTER D. STEPHENS

The Eagle's Nest: Natural History and American Ideas, 1812–1842 by Charlotte M. Porter

Nathaniel Southgate Shaler and the Culture of American Science by David N. Livingstone

Henry William Ravenel, 1814–1887: South Carolina Scientist in the Civil War Era by Tamara Miner Haygood

Making Medical Doctors: Science and Medicine at Vanderbilt Since Flexner by Timothy C. Jacobson

Granville Sharp Pattison: Anatomist and Antagonist, 1791–1851 by Frederick L. M. Pattison

F. L. M. PATTISON

GRANVILLE SHARP PATTISON
Anatomist and Antagonist
1791–1851

THE UNIVERSITY OF ALABAMA PRESS

For sale only in the United States and Canada

Published by
The University of Alabama Press
P.O. Box 2877, Tuscaloosa
Alabama 35487

Library of Congress Cataloging-in-Publication Data
Pattison, Frederick L. M., 1923—
    Granville Sharp Pattison: anatomist and antagonist,
1791–1851.
    (History of American science and technology series)
    Bibliography: p.
    Includes index.
    1. Pattison, Granville Sharp, 1791–1851.
2. Anatomists—Great Britain—Biography. 3. Anatomists—
United States—Biography. I. Title. II. Series.
QM16.P3P38 1987 611'.0092'4 [B]   87–10835

ISBN 0-8173-5154-x (alk. paper)

. . . greatly to find quarrel in a straw when honour's at the stake
—Shakespeare, *Hamlet*

For the curious, pick up the literature on the Chapman-Pattison quarrel, and anything, in fact, relating to that vivacious and pugnacious Scot, Granville Sharp Pattison.
—Sir William Osler, *Aequanimitas*

Contents

Illustrations

Preface

I first became aware of my great-great-great-uncle Granville Sharp Pattison when, as a boy, I heard family members talk in whispers about an ancestor who was clearly considered a black sheep for his activities as an anatomist, body-snatcher, and duellist. It was difficult to uncover details then because, in the words of an elderly Scottish aunt, 'in my younger days, anything derogatory about the older members of the family was carefully concealed from the younger generation'. After becoming a medical student some thirty years later, I decided to hunt out all that I could about this enigmatic and abrasive forebear.

It was not hard to find extensive documentation regarding a few aspects of his life, such as the pamphlet wars with Nathaniel Chapman and William Gibson in Philadelphia. But other episodes required reconstruction from dusty, hand-written documents held in archives on both sides of the Atlantic. I have described the events that emerged at some length, believing that 'generalizations belong to the devil; particulars to the Lord'. Gradually I came to know the man very well—probably better than his most intimate acquaintances.

During my many visits to the five cities in which he worked, I found that he is still remembered with fond interest. Indeed, each city claims him as one of its most memorable characters, if only for the storms he aroused. Through this common bond, I made many new friends and even discovered some distant relatives. They all in turn provided the help and advice so indispensable to a fledgling historical biographer. As such, I encountered all the usual frustrations: a lack of material at some points, an over-abundance of detail at others, the continual problem of balancing background information with the principal subject, and the difficulty of attaining symmetry and flow. Hester Chapman has aptly suggested that the historical biographer often emerges 'as a pitiable and ludicrous figure, stumbling out of one quagmire into another'.

But these limitations did not obscure the general shape of Pattison's career with its many ups and downs. Low points included his crises at Glasgow and London, while the apogee occurred at Baltimore. The vicissitudes of his life were finally resolved when, during his last years in New York, he enjoyed the affection and respect due an elderly professor.

My objective was to study the achievements and adventures of one of Scotland's most colourful nineteenth-century personalities and to use the chronological narrative of his life in Britain and the United States as a framework for examining contemporary social history and international aspects of medical education and practice. Many of the issues are still relevant today. Pattison encountered bitter professional rivalry, particularly in Philadelphia and London; and student activism was so rampant and powerful that he was dismissed without cause from the chair of anatomy at the University of London.

Issues such as these have inspired two plays about Pattison: *Mrs McAllaster's Body* by George Maclean, a forty-five minute radio drama broadcast on 23 January 1957 by the BBC, and *Bedticks and Bloodstock* by James McCash, a period play read by the Glasgow University Arts Theatre Group on 4 June 1973. In the course of my research, I gave a talk entitled 'Uncle Granville' (BBC Radio 4, Scotland, 9 July 1971) and was interviewed about Pattison's role as a body-snatcher on the programme *Good Morning, Scotland* (BBC Radio Scotland, 7 July 1976). These broadcasts and a letter of intent published in the *Scots Magazine* (January 1977) resulted in additional contacts and leads.

It was these contacts and very many other colleagues and friends who provided so much help and encouragement. My warm thanks are extended to the librarians and staff of many institutions in Scotland, England, and the United States, most of which are listed in the text or Notes of this book. I wish to express my gratitude to a number of people who proferred invaluable advice and assistance: Mr Van Afes, Dr Robert Anderson, Mr Jack Baldwin, Dr John Blake, Dr Simon Boniface, Dr George Callcott, Dr Margaret Menzies Campbell, Dr James Dallett, Mrs Kathryn de Mange, Dr Derek Dow, Mr Joe Fisher, Mrs Elvina Foster, Ms M. Fransiszyn, Mrs Ellen Gartrell, Miss Joan Gibbs, Professor James Greig, Mr David Hamilton, Mr John Hamilton, Dr Negley Harte, Mr Donald Hay, Mr James Highgate, Mr Joseph Jensen, Mr Archie Lee, Dr J. B. Lyons, Mr George Maclean, Sister John Mary, Mr Gordon Mestler, Dr J. B. Morrell, Mr Michael Moss, Dr Leslie Murison, Miss Irene O'Brien, Dr Valdis Pakulis, Mr Peter Parker, Miss Eleonor Pasmik, Mr R. M. Price, Mr Ted Ramsey, Dr Alex Rodger, Dr Robert Shanks, Mrs Elspeth Simpson, Dr A. M. Wright Thomson, Mr P. Wade, Mr Ronald Wade, Miss Elizabeth Wilson, and Dr T. E. Woodward.

I am particularly grateful to Dr Leslie R. C. Agnew, Mrs Edith Frame, Mrs Mary Manchester, and Miss Jessie I. M. Wilson for kind and sympathetic guidance in the early days of my work; to Professor David M. Walker for expert advice on a variety of legal matters relevant to Pattison's troubles in Glasgow and for a note about punishments for grave-robbery; and to Professors Eric Atkinson and Richard Shroyer for reading the manuscript and gently helping me correct the worst of my faults. My thanks go too, to Mrs Patricia Squires and Ms Wendy Harris, who kindly provided editorial advice, and to Ms Marion Clarke, Mrs Judy Flannigan, and Mrs Madeline Jillard, who expertly transcribed the several versions of my manuscript.

I gratefully acknowledge generous financial assistance from Associated Medical Services, Inc. and the Hannah Institute for the History of Medicine, Toronto, and wish to record my deep appreciation to the Rockefeller Foundation for enabling me as a scholar-in-residence to write part of the book at their Study Center, Bellagio,

Lake Como, Italy.

Finally, I would like to thank my wife Anne and our four children, Penny, Elizabeth, Rosalind, and David, for their warm interest and support.

F. L. M. Pattison
London, Ontario
1987

Frontispiece: Granville Sharp Pattison, engraved and printed by J. Sartain after the portrait by Chester Harding, 1826. (Royal College of Physicians, London)

I
Early Days in Glasgow
1791–1813

GRANVILLE SHARP PATTISON was a man both admired as an anatomist, surgeon, and teacher, and disdained as a grave-robber, adulterer, duellist, and inveterate wrangler. The dichotomy of outstanding professional attainments and bitter quarrels was to be his hallmark throughout life.

By the age of twenty-seven he had been tried before the High Court of Justiciary in Edinburgh for grave-robbery, found guilty at the Glasgow Royal Infirmary of unprofessional conduct, and named as paramour in the divorce proceedings of a university colleague. But during this time, his ability as an anatomist and his eloquence as a teacher had attracted and inspired the loyalty and devotion of a very large band of Scottish medical students.

Forced into voluntary exile in America in 1819, he was soon embroiled in fresh controversy, this time with medical colleagues at Philadelphia, involving pamphlet wars and culminating in a duel with a general in the United States Army. Yet, within a year of his arrival he had become an esteemed, highly influential faculty member at the University of Maryland, leading a dramatic expansion of its medical school and inaugurating one of the first modern residential teaching hospitals in the United States.

His return to England and election in 1827 as a founding professor at the newly constituted University of London ended in his dismissal after four quarrelsome years. Nevertheless, his numerous supporters, no less vociferous than his detractors, proclaimed their admiration for Pattison as a gentleman and a teacher.

A Child of His Time

Who was this extraordinary man? Like all of us, he was a child of his time. When he was born, the French Revolution had just begun; he died in the year of the Great Exhibition at the Crystal Palace in London, which ushered in the Victorian era. He belongs, then, chronologically to the age of Romanticism: the age of Hegel and Schopenhauer; of Delacroix, Turner, and Constable; of Beethoven, Schubert, and Chopin; and of Blake, Wordsworth, Shelley, Scott, and Carlyle. This period was in many respects the beginning of the modern era. Thoughtful men were already seeking to resolve the fundamental social problems which still confront us. And in the work of its artists and philosophers may be found the roots of modern psychology.

Like Pattison himself, the social background of his formative years presents striking disparities. Glasgow at the end of the eighteenth century, a rural town of 50,000 inhabitants, was a mixture of affluence and poverty, tranquillity and violence, intellectual pretension and widespread ignorance.

The affluent enjoyed comfortable circumstances. As a boy, Pattison, son of a wealthy Glasgow merchant, lived in an eight-bedroom mansion on a large estate with a full staff of servants and a well-stocked larder; and he attended the only private school in Glasgow. In contrast, the mass of the poor lived in crowded slums of unspeakable squalor, where they subsisted largely on potatoes and oatmeal. Their children received little or no education; many under the age of nine worked a minimum of forty-eight hours a week in factories, and children under five were commonly employed in coal mines. The young Pattison could not have failed to have been affected by the advantages of his family's social position in comparison with the conditions of the poor. Such circumstances may well have contributed to the arrogance he often displayed in later life.

Another social difference must have influenced the life of this complex man. In Scotland no less than England, the respectable upper middle class to which the Pattisons belonged enjoyed that ordered existence portrayed by Jane Austen, Anthony Trollope and other English nineteenth-century novelists. Contrasting with this tranquillity was the violence of the rebellious poor. Riots were frequent and difficult to control, since policing was normally left to the citizenry, assisted by a small, armed constabulary. Those convicted were brutally punished. Pattison himself was to taste both mob violence and the disgrace of the judicial process when he was arrested and tried for grave-robbery. Indeed, throughout his life he seems almost to have courted violence.

Medical science displayed other contradictions. Men of widely differing motives and abilities worked to alleviate that human suffering which most accepted as 'the will of God'. Two out of every five children died before the age of two, and about half before their tenth birthday. Of these deaths, one-third was attributable to smallpox. When the prevalence of smallpox declined with the introduction of vaccination, more children survived, only to succumb a few years later to measles, diphtheria, or tuberculosis. Sanitation was unknown, and epidemics ravaged the population. Some practitioners, clinging to ancient dogma and the pretence of wisdom, propounded ludicrous solutions. Others, with some perception of the enormous gap between their knowledge and the problems they faced, strove valiantly to overcome the primitive state of general medical management, based as it was on a mixture of superstition and gross empiricism. Here was a situation which might well have prompted a man of Pattison's temperament to accept the challenge of a career in medical teaching and research.

Early Childhood

It was into this world of social extremes and medical incongruities that Pattison was born. In the Glasgow Barony parish records, in the county of Lanark in Scotland, there appears the following entry:

January 1791: John Pattison, Esq., of Kelvin Grove and Hope Margt. Montcrieff [*sic*] had a lawful son, born 23d, bapd [blank], named Granville Sharp. Geo. Bell and Thos. Mitchell, wit.[1]

John Pattison of Kelvingrove was a prosperous Glasgow merchant and muslin manufacturer; his wife, Hope Margaret Moncrieff, came from an ancient and highly respected Scottish family.

They named their son after Granville Sharp, the renowned opponent of the slave trade. While the parents showed courage in naming their son after such a zealous abolitionist and supporter of the American revolutionaries, it may well have been hard on the boy to be linked by name with a cause that was so violently opposed by many of the wealthy and influential members of the community. There were family reasons as well as personal convictions that associated the Pattisons with the abolition of slavery. John Pattison had a brother-in-law who was a friend of John Newton, the distinguished hymn-writer, erstwhile slaver, and later outspoken adversary of slavery. And the Pattisons are thought to have known Dr John Jamieson, the Scottish lexicographer and author of the poem *The Sorrows of Slavery*, whose abhorrence of the slave trade is well known.

Besides fighting slavery, Granville Sharp violently opposed the press-gangs that shanghaied able-bodied men for service in His Majesty's Royal Navy. Frequently the press-gangs would board merchant ships and carry off the crew, often within sight of the home port. Since much of John Pattison's business depended on the export of his muslin to the continent, the press-gangs' activities among the merchant fleet must have been a constant source of anxiety to him—and another reason for admiring Granville Sharp.

Granville Sharp Pattison was the sixth of eight surviving children, the fifth surviving and youngest son. Two brothers, both born before Granville, died in early childhood; two others, John and Alexander Hope, survived to play a part in his troubles in the United States. He had three sisters, including Margaret who was born next after him. She was involved in several episodes in his later life and was a beneficiary under his will. A number of their relatives were notable figures of the day. They included Liberal MPs, Fellows of the Royal Society, and Lord Kelvin, the great Scottish physicist and mathematician, a first cousin once removed. Further details of the Pattison family are given in Appendix 2.

There is little specific information about Pattison's early upbringing, but we may assume that he enjoyed the comforts and claims of his class. It was customary then for prosperous families to have nursemaids and a nanny or governess to look after young children and to provide a modest degree of basic education (often accompanied by a more than modest degree of discipline). Occasionally the nursery staff was augmented by a tutor. Parents rarely encountered their young children except for a brief evening session, after which the children were removed by the nanny or governess to allow the parents to proceed to dinner. For this rather formal meeting, the children were meticulously washed and groomed. It was, in short, a regimen that did not encourage intimacy between children and parents, especially fathers. In later life Pattison was to show an affection for his mother, but such few references as he made to his father were cool.

When he was a year old, the Pattisons moved to Kelvingrove House, their home from 1792 to 1806. Like most mansions of the time, it was without what we now consider to be essential sanitary arrangements, but its elegance and luxury were in stark contrast to the sordid conditions existing for most of the citizens of Glasgow. The house, situated on the River Kelvin, one mile from the city and built by Robert Adam, lay in twenty-eight acres of gardens and woods. It was described in a newspaper advertisement placed by the original owner, Patrick Colquhoun, in 1790: 'The house, which overlooks the river, is built on a very commodious plan, containing a dining-room, drawing-room, eight bed-rooms, two lumber-rooms, a kitchen, larder, and three cellars under ground. The offices consist of a stable, with stalls for four horses, a cow-house, milk-house, chaise and cart house, a hay-loft, pigeon-house, poultry-houses, etc., all in the most complete repair. There are also a pump-well in the yard, a convenient wash-house, with a pipe from the river, and a large and commodious cold bath.'[2]

John Pattison of Kelvingrove, by Sir Henry Raeburn. From the catalogue of the 1868 Glasgow Exhibition of Portraits. (Mitchell Library, Glasgow)

Hope Margaret Pattison née Moncrieff, by Sir Henry Raeburn. From the catalogue of the 1868 Glasgow Exhibition of Portraits. (Mitchell Library, Glasgow)

The property had an extensive garden surrounded by a brick wall ('part of which has flues'), woods, and a large variety of flowering shrubs and fruit trees. The beauty of the estate was extolled in a contemporary ballad 'Kelvin Grove' by Thomas Lyle:

> O Kelvin banks are fair, bonnie lassie, O,
> When in summer we are there, bonnie lassie, O,
>   There, the May-pink's crimson plume,
>   Throws a soft, but sweet perfume,
> Round the yellow banks of broom, bonnie lassie, O.[3]

John Pattison undertook extensive improvements to the grounds. Near the house he planted a series of Canadian poplars to commemorate the birth of each successive member of his family. In later years the trees, one associated with Granville, came to be known as 'the Pattisons'.

It is likely that the Pattison boys swam in the River Kelvin, which flowed gently through the estate. A writer in 1872 recalled that in his youth boys used to bathe in the Kelvin, adding that since then it had become a 'filthy element, stinking and stagnating'.[4] Kelvingrove, with its stables and extensive grounds, also provided them with the opportunity to ride. In later life Pattison enjoyed fox-hunting.

Family tradition has it that John Pattison, on attaining the station of a laird, over-extended himself in improving Kelvingrove. At any rate, his business affairs deteriorated to the point where he was forced to sell the whole property in 1806.[5] After his death the following year, the proceeds of the sale were used to settle outstanding debts, to purchase commissions in the army for his sons Alexander Hope and Frederick, and to complete Granville's medical education.

Kelvingrove undoubtedly had an influence on the development of Pattison's character. Its luxury and social pre-eminence would not have been lost on him. But of greater consequence was his father's inability through loss of fortune to retain the property. Its sale in 1806 marked the family's slide from affluence into relative poverty; and the move from Kelvingrove to 2 Carlton Place, a terraced house in the centre of Glasgow, must have been traumatic for a fifteen-year-old boy.

Kelvingrove House. From *The Old Country Houses of the Old Glasgow Gentry*. (Mitchell Library, Glasgow)

It is perhaps significant that the move occurred during the hiatus between his schooldays, which were anything but auspicious, and his successful medical training. The protective splendour of Kelvingrove may have contributed to a slothful performance at the Glasgow Grammar School; stripped of the security associated with a wealthy father and a fashionable address, he would surely have perceived the urgent need for sound training and a productive career.

Glasgow Grammar School

In 1802, at the age of eleven, Pattison became a pupil at the Glasgow Grammar School, where he remained for a period of only two and a half years.[6] He was a very poor student; his class records bear witness to that. In view of his later outstanding achievements, it would seem that he was a slow starter, a type familiar to every experienced teacher: not dull-witted, but obstinate in his refusal to learn what did not interest him. Only later, through inspired teaching, was he to be animated by the discovery of his vocation.

The winter hours of the school were 9.00 to 11.00 a.m., and noon to 2.00 p.m., while the summer hours were 7.00 to 9.00 a.m., 10.00 a.m. to noon, and 1.00 to 3.00 p.m. A four-week vacation began on 1 July, and there were in addition a large number of 'play days'—religious breaks, May Day, and the King's birthday. Each boy paid five shillings per quarter for tuition, and sixpence per quarter for coals.

The business of the school was conducted by four masters of equal rank, authority, and salary, each of whom in turn began an elementary class and carried it forward for four years. The masters took turns in serving for a year as rector, which meant presiding at all school gatherings, directing and implementing discipline, and saying the morning prayer each day.

The masters, usually in absolute control, adopted an obsequious role on one day in the year, when the boys presented them with a gratuity or 'Candlemas Offering'. For this occasion, on 2 February, the boys were convened in the common hall. When the masters were seated at their desks, the boys in all the classes were expected to walk up, one by one, to their master, and give him an 'offering'. Some boys showed their contempt for the proceedings by paying their offerings in coppers (pennies or less) in order to enjoy the sight of the masters counting their presents.[7]

Latin was studied extensively in all four years, with Ruddiman's *Rudiments* and Mair's *Introduction* providing the basic grammar and syntax. Greek was studied in the third and fourth years only, for which Moor's *Greek Grammar by Tate* was the standard textbook. 'General Knowledge' was included in all four years; this was a catch-all course encompassing British, Roman, and Greek history, English grammar, composition, and literature, and geography, but not science or mathematics. The Bible was read in every class, and select passages were frequently committed to memory. This syllabus differed little from that offered by other British grammar schools, and the training conferred on Pattison the standard middle-class education of the period. The scripture lessons seem to have made little or no impression on him; his writings show little interest in religion.

Glasgow Grammar School. From *Glasghu Facies*. (Mitchell Library, Glasgow)

The most junior class was referred to as the fourth (*classis quarta*), with subsequent enrolment *in classe tertia*, *in classe secunda* and finally *in classe prima*. Opposite each pupil's name in the class books are his numerical position in the class for each of the eight or nine tests held during the year, and their sum; his final standing in the class at the end of the year (if he were one of the better students); and any academic prizes won. On the last pages of each year are summarized the number of boys in the class, the number of academic prizes awarded, the names of the recipients of attendance prizes, and a note about the actual topics which were covered in each of the individual tests.

Pattison's name first appeared on 5 May 1802 as a pupil *in classe quarta*. When he was *in classe secunda* (1803/04), there was uncertainty about the possibility of completing the school session because of the threat of imminent invasion by Napoleon Bonaparte. Throughout his time at the Glasgow Grammar School, he was taught by David Allison, an experienced schoolmaster in his early fifties, in a class of 100–120 boys.[8]

Pattison, invariably listed as 'Grenville Pattison', was always in the bottom half of the class.[9] He failed to receive any of the thirty-four academic prizes awarded in his class each year. Even his attendance was erratic. Only *in classe tertia* (1802/03) did he receive an attendance prize (one of fifty-eight): a copy of Oliver Goldsmith's *History of England* (abridged). This seems to have been the only occasion that Pattison was eligible for one of the prizes presented by the Lord Provost and Magistrates at the prize-giving ceremonies held annually at the end of September. His last year's record shows that he wrote but one test, on 29 November 1804. He answered questions on the writings of Virgil, the first forty rules of prosody, and parts of Ruddiman's *Rudiments* and Mair's *Introduction*. He was eighty-seventh in a class of 104 boys. In the class records David Allison wrote the single bleak word 'gone'.

These school records provide the first sign of a certain indolence which was to be apparent throughout his life. But his poor grades may suggest as well that he was out of tune with his own cultural heritage. Latin at that time occupied a large place in vernacular Scots and was as familiar to the educated Scot as his own mother tongue. It was the main subject in all four years at the Glasgow Grammar School, but it obviously did not appeal to Pattison. Indeed, his ignorance in Latin was later emphasized in the accusations made against him during the last and most bitter phase of the London hostilities: the students there charged him with 'a want of a liberal education, frequently evinced by the commission of flagrant classical errors', and supported their charges by quoting Latin grammatical errors allegedly perpetrated by Pattison.

The University of Glasgow

Given Pattison's undistinguished performance at school, it seems likely that his father withdrew him in November 1804 to provide him with private tuition over the next two years. It was during this time that his family became *déclassé* through the loss of Kelvingrove. Expediency may then have demanded that he follow a medical career. Later, when he came under the influence of two of Glasgow's most renowned anatomists, James Jeffray and Allan

Burns, he found that anatomy was his true vocation.

He attended the University of Glasgow in the period 1806–12, matriculating in 1807 in his second year of study. The formal registration in his autograph appears in the Matriculation Album: 'Granville Sharp Pattison, *f.n. ʃtus Joannis, Mercatoris,* Glasg. Lanark.'[10] He did not take a degree, so his name is not included in the corresponding Graduation Album. At this time, however, students rarely went through the formality of graduating; certificates of attendance at the different classes provided sufficient credentials for medical or surgical practice.

The University of Glasgow, founded in 1451, had provided instruction on medical topics since the early years of the eighteenth century.[11] As the majority of the medical students belonged to the middle or lower classes, Pattison's decision to enrol in medicine may have caused his parents some concern. Medicine, however, was one of the few suitable careers open to an impecunious youth of modest academic attainments, particularly as there were no entrance requirements. This situation differs greatly from the high academic standards and large fees demanded of prospective medical students in many countries today.

At fifteen Pattison was older than most students entering medical school. The number in medical classes averaged about 200 but varied greatly according to the demand for military surgeons; the classes were large in Pattison's day because of the Napoleonic Wars.

At this time the Glasgow Royal Infirmary was run by the powerful Faculty of Physicians and Surgeons of Glasgow, the members of which had the sole authority to license medical practitioners in Glasgow. The university professors resented their inability to license their own graduates; intense rivalry prevailed. As a result, all the professors were barred from hospital practice and bedside teaching. It was in this stormy professional climate that Pattison's medical studies began. Not many years later he himself was to become the centre of fresh controversies and quarrels in Glasgow.

Pattison attended a number of classes: 1806/07 Humanities (i.e., Latin); 1807/08 Greek; 1808/09 Anatomy; 1809/10 Practice of Medicine, Chemistry, Botany, Infirmary; 1810/11 Clinical lectures, Infirmary; 1811/12 Materia Medica.[12]

Old College of Glasgow. From *Glasgow Illustrated in Twenty One Views*.
(Mitchell Library, Glasgow)

His pre-medical courses, both of one academic year's duration,
provided twenty-three hours of instruction per week. Latin included the language, literature, history, and antiquities of ancient
Rome, while in Greek the students concentrated on grammar, verse,
and prose. During subsequent years, Pattison studied two other
non-clinical subjects: Chemistry involved the study of inorganic
substances and the analysis of minerals; Botany, taught by a deputy
of James Jeffray, the professor of anatomy, included taxonomy and
the anatomy and physiology of plants, and was augmented by field
trips throughout the year.

Of the standard medical courses followed today, two—surgery and
midwifery—are, at first glance, lacking from Pattison's university
curriculum. Surgery was then included as part of the anatomy course,
while midwifery—encompassing obstetrics, gynaecology and paediatrics—was taught privately by John Burns at the College Street
Medical School, with which Pattison became closely associated.[13]

In anatomy, Pattison's first medical course, he found an inspiring teacher within a system that was at times seriously defective. Lectures were delivered by the distinguished professor, James Jeffray, whose fifty-eight-year occupancy of the chair at Glasgow (1790–1848) is a record for tenure in a medical chair at any Scottish university. His lectures, delivered with dignity, finesse, and contrived artistry, held his students spellbound. He was at least partially responsible for Pattison's lifelong espousal of anatomy.

As well as electrifying students with his oratory, Jeffray was one of the first to electrify a corpse. When Pattison was still working in Glasgow, Jeffray and Andrew Ure investigated the effects of a new and very powerful battery on the body of a recently executed murderer, William Clydesdale. Stimulation of different nerves produced dramatic results, such as movements of limbs, smiles and grimaces, and laborious breathing.

Jeffray gave a one-hour lecture each weekday and a review examination each Saturday. The syllabus, fairly standard for anatomy courses of the period in both Britain and the United States, covered in turn bones, muscles, heart and blood vessels, the absorbents,[15] the brain and nerves, the viscera, physiology, and pathology.

After his lecture Jeffray's formal teaching duties for the day were over. But the students were encouraged to work in the dissecting room for a further four hours and to attend a second daily set of lectures, given this time by the demonstrator. Such a seemingly satisfactory arrangement was sometimes marred by the incompetence of the demonstrator. Indeed, perhaps the worst feature of the teaching of anatomy was the professor's lack of supervision of the dissecting rooms. The daily activities in 1812 were described by Thomas Lyle who attended classes between 1812 and 1815: 'The dissecting room was well attended at this time, and the dissector, Mr Jardine, drew a great deal of money from the students, we are sorry to say, without making them any adequate returns. The herd of students who lounged there derived little or no information; their time was passed running out and in—like wild and idle schoolboys —as they pleased, without being kept under any control or discipline whatever. . . . This place that ought to have been of the utmost importance to the student in a practical point of view, was neither in the face; another was found dead after being thrown over the parapet of a lofty bridge.

Pattison's first encounter with anatomy showed him some of the best and worst features of its practitioners. His ultimate decision to pursue it as a career suggests that Jeffray's skill and eloquence outweighed any incompetence of the demonstrator. He himself was never happier than when instructing students in the dissecting room or when lecturing to a large and appreciative audience. It was when these activities were denied him through the animosity or incompetence of others that he became truculent and aggressive.

Pattison attended lectures in the practice of medicine (earlier known as 'theory and practice of physic') given by Professor Robert Freer the following year (1809/10). Students are quick to assess the teaching ability of their professors. Lyle's regard for Jeffray did not carry over to Freer, who 'never was, we are sorry to say, much esteemed as a popular lecturer, his dogmas and doctrines being deemed stale and uninteresting, and might have passed current about a century ago, but certainly his lectures were by no means adapted to fascinate the attention of the student . . . who expected he would have attended more to the changes and improvements progressively taking place in medicine at the present day'.

Freer's class met for one hour each weekday, according to the University Calendar, 'when the various diseases which come under the attention of the physician are treated of nearly in the order in which they are arranged in Dr. Cullen's *Nosology*'. A related class on the theory of medicine (institutes of medicine) met for a further three hours each week, 'when the different subjects connected with it are treated of nearly in the order in which they are found in Dr. Gregory's *Conspectus Medicinae Theoreticae*'. The students were examined publicly three days a week on the lecture material.

Dr Richard Millar lectured on materia medica during the academic year 1811/12. The course, which occupied one hour each weekday during the winter, consisted of chemical and Galenical pharmacy and dietetics. This included the sources, preparations and uses of drugs and potions. The requisite knowledge of practical pharmacy was gained by working for a given number of hours daily during three months in a recognized druggist's shop. Pattison and

more nor less than a den of tumult and uproar, of wrangling and brawling and every species of schoolboy mischief when the master is absent, during the whole forenoon.' Occasionally Jardine would punish the students for fighting each other with loose bones by locking up all the specimens, thereby penalizing the industrious student along with the guilty.

The contrast between the dignified Jeffray and his irresponsible demonstrator points up the remoteness from their students of even the most inspiring professors of this period. Jeffray was content to give his polished lectures, but remained aloof from the day-to-day activities of the dissecting room. (He was a capable dissector; the anatomy museum at the University of Glasgow still includes some of his preparations.)

The demonstrator's irresponsibility was not the only trouble encountered by the students. Even more significant were the hazards associated with dissection itself. A physician wrote in 1794: 'Within the last year, five lecturers in anatomy are now fresh in my memory, who have fallen the victims of putrid miasma in the prime of life.'[16] And the threat of physical violence was always present. Condemned criminals, the only legal source of specimens in Scotland, feared and detested the thought of being dissected by the doctors. The friends of a man hanged in 1820 were so outraged at his dissection that they took summary revenge. All the medical men who had taken part received personal injuries. One doctor was shot

James Jeffray (*opposite*), professor of anatomy at the University of Glasgow, engraved by Edward Burton after the painting by John Graham Gilbert. From the catalogue of the 1868 Glasgow Exhibition of Portraits. (Mitchell Library, Glasgow)
'Whenever he raises himself up from his tripod, and emerges from behind the revolving mahogany dissecting-table to the middle of the arena fronting the students—takes off his glasses—wipes his eyes with his snow-white handkerchief—plants his right foot a little in advance—draws up his tall and stately form to its full height, and slightly separates his quivering lips after the full inspiration is slowly expired again—then it seems that a more than usual silence pervades the whole theatre, the student becoming aware that these are preliminaries of something grand about to follow . . . .
Imperceptibly the professor, as well as his audience, are completely electrified and as it were spellbound within this melody of eloquence, and their senses lost in the intensity of the subject before them . . . . His audience are left nonplussed whether the orator, or his subject, are to be admired most.'[14] (Thomas Lyle)

Millar were to meet again some four and a half years later when the latter chaired a committee of enquiry examining accusations against his former student of unprofessional conduct at the Glasgow Royal Infirmary.

Pattison's studies at the University of Glasgow, then, provided him with most of the basic medical training at a fairly leisurely pace over six years. He probably attended some of the meetings of various societies, such as the recently founded University Medico-Chirurgical Society, where students gathered to discuss the contents and implications of the lectures that they had attended. These encounters, which provided the all-important stimulation and speculation apparent in a well-conducted tutorial today, may well have spawned his fondness for debate.

By the end of his university studies, Pattison was a self-confident young dandy of twenty-one. At about this time he was caricatured as 'Beau Fribble' in a vitriolic little book entitled *Northern Sketches or Characters of Gxxxxxx* (Glasgow) by a pseudonymous author, 'Leonard Smith'. The satirical observations are heavy-handed: '[Beau Fribble] acted for a little time the part of a fool, and then attempting the character of a fine gentleman, fell, from a love of his native country I suppose, into that of an ape. In this shape I have presumed to exhibit him—acting like an ape, grinning and prating like an ape, and biting also like an ape.'[17] But even this kind of writing might have been less distasteful to Pattison than Peter Mackenzie's patronizing description of him as 'a bold, clever Glasgow youth'.

Whatever the truth about his appearance and reputation, Pattison throughout his life was invariably described as kind and generous. One of his young colleagues in Glasgow wrote, 'I am under too many obligations to [Pattison] for his former kindness to advise anything likely to be injurious to him.'[18] Little is known about his political and religious affiliations. The Pattison family was held to have traditional Liberal or Reform tendencies. It is likely that they worshipped at the Ramshorn Church, as John Pattison was buried in its graveyard. Another member of the Ramshorn congregation was a Mrs Janet McAllaster, who was to play a notable role in Pattison's professional life. In his leisure hours, he enjoyed

Old Ramshorn Church. From
Glasghu Facies. (Mitchell Library,
Glasgow)

College Street Medical School, the
second building from the left, with
the low roof. From *Glasgow
Illustrated in Twenty One Views*.
(Mitchell Library, Glasgow)

attending a variety of plays and operas, which were becoming more
available through the gradual easing of Calvinistic repression. But it
was his medical career that was occupying more and more of his
time.

College Street Medical School

Medical education in Glasgow was not limited to the university. It
was in 1809, while still attending some classes there, that Pattison's
association with the College Street Medical School began.

This thriving private establishment, founded in the late eighteenth century, had no formal links with the university; it catered to
all sorts of people including university students, surgeons and
apothecaries in training, and dilettante men-about-town. The

competition between private schools and the university certainly gave impetus to learning, and students were often encouraged to attend courses at both. The date of the school's closure is not known; lecturers were still being appointed in 1835.

In 1809 the school moved to premises at 10 College Street, on the north side and close to the east end.[19] Many of its students conveniently lived in the upper flats on each side of the street. Some of them stretched double lines of cord from window to window, to transfer small articles across the street from one lodging to another. The general impression from various contemporary accounts is one of bustle and activity, typical of a thriving student community.

The school's location allowed easy access by a narrow passage, called Inkle Factory Lane,[20] to the nearby Ramshorn Churchyard*—and to a plentiful supply of cadavers. The students, supported by their teachers, ensured that there were always subjects for dissection. The school's unofficial status provided some immunity from investigation. The university, on the other hand, with its stricter code of ethics, demanded greater circumspection in the practical aspects of anatomy and surgery.

No records of any kind survive from the College Street Medical School, but fragments of information are available from contemporary letters. One of these states that when Pattison lectured in the 1818/19 session, he had eighty-six students, each of whom paid three guineas; and another that the rent for his rooms that year was thirty-six pounds per annum. The teachers were conscripted from among the local physicians, surgeons, and anatomists. Clearly, education at this time was dominated by individuals rather than institutions, and the College Street school had its full share of memorable personalities.

Foremost among these were the Burns brothers, John and Allan. John founded the school, but it was his younger brother Allan who had such a profound effect on Pattison's professional development. This brilliant young man, continuing what Jeffray had begun, was instrumental in guiding Pattison toward his lifelong study of anatomy. He later remarked that he was scarcely separated from

*See p. 32 for a street plan of the area.

Allan Burns. (Royal College of
Physicians, London)
'He was small, delicate and boyish
looking, and the first time we saw him
enter the theatre to deliver his
introductory lecture surrounded by
several of the private party, who were
stout robust fellows, we mistook him
for one of the apprentices, till he
emerged from the crowd, advanced to
the table, and commenced reading his
lecture.' (Thomas Lyle)

Allan Burns for as much as an hour, and his friendship with him was
more like that of a brother than a preceptor. This association was
pivotal in Pattison's life.

In 1797, the sixteen-year-old Allan Burns was in charge of the
school's dissecting rooms, while his brother John lectured on
anatomy. Allan developed outstanding skills in dissection and in
devising new methods of specimen preservation, paying special
attention to the vascular system. Having no Glasgow qualification,
he was debarred from surgical practice; instead, he visited a number
of his brother's patients and, if they died, attempted to correlate
their symptoms with the post mortem findings.

By 1804 Allan Burns's fame was such that he was invited by the
Empress Catherine to direct a hospital along British lines in St
Petersburg and to serve as physician to the Imperial Court of Russia.
He accepted the invitation for one year. On his return in 1805 he

learned that John had been prosecuted and barred from teaching anatomy because of his proved involvement in a case of grave-robbery. This provided the opportunity for Allan to take over the anatomy class. He gave lectures on anatomy and the principles and operations of surgery; the course fee was two and a half guineas.

Allan Burns's style of composition was denigrated as too diffuse and colloquial. But he was a good lecturer, and his boundless energy, enthusiasm and skill attracted a group of young men inspired by his love of anatomy. Pattison allowed that Burns could be irascible, but added that 'his passion was but for a moment; and if, under its influence, he did anyone an injury, he was the first to confess it and make ample reparation'.[21]

In 1809, at the age of eighteen, Pattison became Allan Burns's assistant and demonstrator and worked with him until Burns's death on 2 June 1813. Characteristically, Pattison's response to his mentor's death was scientific as well as sentimental. Burns had suffered from an extremely painful abdominal abscess, which had burst into the rectum, discharging large quantities of pus. These pathological details were quickly determined by Pattison and two colleagues, who had performed an autopsy on their friend immediately after his expiry.

Burns has been immortalized in two eponyms—'Burns's ligament' and 'space of Burns'[22]—and two renowned books, *Surgical Anatomy of the Head and Neck* and *Observations on Diseases of the Heart*. He bequeathed to Pattison the copyright of all of his works, the best known of which Pattison revised and expanded to include a biography of Burns and a large number of additional clinical cases.

During Allan Burns's years in Glasgow, he and his senior associate Andrew Russel, one of Glasgow's town surgeons, co-operated in the preparation of an extensive museum of anatomical and pathological specimens. No record exists of Pattison's contribution, if any, to this collection. Following Burns's death, Russel, to whom Burns had bequeathed his share of the museum, became its sole proprietor but consented to share ownership with Pattison. In a formal contract of partnership, it was agreed that the museum be evaluated and that Pattison be offered the opportunity to purchase a one-third interest. He accepted the offer, and thereupon became the minor partner in the 'firm of Russel and Pattison', drawing one-third of the profits and incurring one-third of the risk. Pattison later acquired title to the museum, which he used to advantage in Glasgow, Philadelphia, and Baltimore.

With the Burns museum in their possession, Pattison and Russel together offered a course of lectures at the College Street Medical School, on anatomy and surgery respectively, for the fee of two and a half guineas.[23] Thus, by the age of twenty-two Pattison had taken up his first teaching position. This was not particularly precocious: John Burns had taught anatomy at the same age.

Thomas Lyle was one of the first to record Pattison's brilliance as a teacher, evident even in these very early days of his career: 'Mr. Pattison's popularity from the time he commenced lecturing was little inferior to that of his late master [Allan Burns]; while he was idolized by the students from the great care and attention he bestowed in grounding them in the principles of anatomy. He must have been a dunce who did not improve under such tuition.'

Not only was Pattison teaching; in 1813 he was also performing major surgery. At this time surgeons generally espoused the Hippo-cratic aphorism that wounds of the belly are fatal. Pattison, challenging this time-honoured belief, claimed to be the first sur-geon in Europe to recommend abdominal surgery for the removal of tumours. (He failed to note that Robert Houston, a Glasgow surgeon, had reported a successful abdominal ovariotomy to the Royal Society in 1724.) Pattison described two cases to the *Société Médicale d'Émulation* in Paris, one of which involved a moribund pregnant twenty-three-year-old woman with an abdominal mass. Pattison, in a twenty-minute operation, performed an exploratory laparotomy and diagnosed a hydatidiform mole, which he treated by scraping off the vesicular villi with his finger-nail. The patient recovered completely and subsequently had two normal deliveries. The French surgeons commended him for the successful outcome, but censured his over-readiness to operate, particularly when the indications for surgery were questionable.

As well as lecturing and operating, Pattison applied for member-ship in the Faculty of Physicians and Surgeons of Glasgow. Three entries in the Faculty Minute Book report his success. On 2 August

1813 his name—invariably recorded as 'Mr Grenville Pattison'[24]—
was proposed by Dr John Balmanno, the president. Five weeks later
he was reported to have passed the required examinations in
anatomy, surgery, and pharmacy. Finally, on 4 October he was
unanimously admitted a Member of Faculty after having read a
prescribed discourse (on crural hernia), exhibited a 'specimen of
medicine' (which was judged by Hugh Miller, his future adversary
at the Glasgow Royal Infirmary), and taken the oaths of membership. He thus became a member of the only body legally competent,
under a Royal Charter from James VI, to license medical practitioners in Glasgow.

   While he was being so honoured, a Mrs McAllaster, the wife of a
Glasgow wool merchant, was ailing. Despite the ministrations of
her physician and friends, she died on 8 December at her house in
Great Hamilton Street and was buried on Monday, 13 December
1813 in the Ramshorn Churchyard. In her life a respectable woman,
Mrs McAllaster in her grave was to occasion the first of the public
scandals which marred Granville Sharp Pattison's professional
career.

II
The Resurrectionist Trial
1813–1814

AT THE TIME Pattison began lecturing at the College Street Medical School, competent surgeons were in great demand for service in the prolonged Napoleonic Wars in Europe. Their training in human anatomy required an ever-increasing supply of cadavers.

Although unclaimed bodies in hospitals and workhouses could be used for dissection in most European countries, in Scotland and England victims of the gallows were the only legal source of specimens. In earlier days criminals had had good reason to fear the anatomist. According to Celcus, renowned physicians such as Herophilus and Erasistratus (*c.* 300 B.C.) felt that human vivisection was especially valuable in determining the workings of the body, and acquired their living subjects by royal permission from prisons. 'It is by no means cruel, by the torture of a few guilty, to search after remedies for the whole innocent race of mankind.'[1] Such in-humanity gradually ameliorated over the centuries until Pattison's day, when condemned murderers alone had reason to dread the anatomist's knife.

Execution by hanging in those days involved slow strangulation. There was no sudden drop to bring about immediate death; instead, the victim mounted a ladder which was then kicked away by the hangman. Occasionally a bucket was used in place of the ladder; friends of the victim sometimes pulled violently on his legs to shorten his sufferings. (These two practices gave rise to the expressions 'to kick the bucket' and 'to pull someone's leg'.) All condemned criminals were haunted by the fear that they would still be alive when the anatomists began their dissection. They did not consider death by hanging to be any great disgrace; that was nothing compared to the indignity of being dissected.

Illegal Exhumation

The legal exclusion of all other sources of cadavers inevitably resulted in a dearth of bodies in Scotland and England. This was occasionally relieved by illicit trade with Ireland, where graveyards were less closely guarded. It is related that an Irish vessel arrived at the Broomielaw (Glasgow docks) with a consignment of bags, listed as 'rags'. The huckster to whom they were addressed refused to take delivery because of the very high freight charges, with the result that the bags were returned to the Broomielaw, where they lay in a shed for a few days. 'An awful stench soon arose from them, and some of the officers, on opening a few of the bags, were horrified to find the ghastly, dead, putrified bodies of men, women and children, huddled together in the most shocking manner.' It seems that the rag merchant did not receive notification of the valuable nature of the cargo until it was too late. The bodies were duly buried, and the rag merchant lost the 'goodly commission for the traffic entrusted to his care'.[2]

Such adventitious sources were inadequate for the growing demand for corpses in the universities and private medical schools. Body-snatching became inevitable. This was highly organized and lucrative, with prices varying from two pounds for a foetus to twenty pounds for a 'healthy adult'. Purchasers occasionally had to pay the grave-robbers for their silence ('protection money') under the threat of having portions of decomposing limbs litter the street outside their dissecting rooms. To guard against grave-robbery, armed sentries were stationed in the churchyards of the rich, while the poor could hire heavy cast-iron slabs and mortsafes to cover the remains of their relatives and friends until time had rendered them invulnerable to the dissector's knife.

The story of grave-robbery has been told many times. Almost all books dealing with the history of medicine devote many pages to the skill, ingenuity, and daring of the 'resurrectionists',[3] the prices obtained for subjects, and the extremes to which an angry populace was forced to ensure the protection of recently deceased relatives. But until the passage of the Anatomy Act in 1832 almost all anatomists and surgeons resorted to illegal means of acquiring specimens. Some of the most renowned openly confessed their guilt.

Pattison himself testified on the practice of grave-robbery before the House of Commons Select Committee on Anatomy, when he was professor of anatomy at the University of London. He considered that a surgeon should have dissected a total of twelve subjects during his medical education. Freely admitting the exhumations he had organized at the College Street Medical School, he explained that every public teacher had what he called his 'private party': this consisted generally of eight students 'and these young gentlemen went out themselves and exhumed the body'.[4] He noted that in Glasgow in 1828 bodies were so scarce that they were salted in the summer, hung up and dried 'like Yarmouth herrings', and the next winter reconstituted by immersion in water.[5]

In Glasgow, body-snatchers were not, as in Edinburgh, the scum of the city's underworld, the 'sack-'em-up' men like Burke and Hare, who committed murder in order to sell the bodies of their victims to the schools of anatomy. (In nine months, Burke and Hare murdered sixteen people in Edinburgh and sold their corpses to Dr Robert Knox, the anatomist.) The Glasgow medical students themselves were prepared to brave mantraps, spring-guns, the forces of law, and the universal execration of the public. They were young and enthusiastic, and carried out their grave-robberies with dash and aplomb, often showing themselves conspicuously in the most frequented taverns on the night of an expedition. They were particularly interested in obtaining bodies of patients with obscure or incurable diseases, and for that purpose had an intelligence service covering a wide area around Glasgow. When news of a suitable specimen arrived, they drew lots for their gruesome task. No protection was provided by the authorities of the medical schools. The nearest approach to official cognizance was the provision of free tickets to the dissecting rooms for those who took part in the exhumations.

Originally a member of Allan Burns's group of grave-robbing assistants, Pattison later became the acknowledged leader of the student resurrectionists in Glasgow, who often displayed a remarkable agility of wit. On one occasion two students exhumed a body in an outlying area and had the problem of bringing it past a toll-keeper, who was known to have a particular aversion to grave-robbery. They dressed the body in an old suit of clothes and sat it between them as they approached the gate. While one paid the toll, the other held up the head of his 'friend', telling him to be of good cheer as they would soon be having their breakfast. The toll-keeper exclaimed: 'O! puir auld bodie, he looks unco ill in the face; drive cannily hame, lads, drive cannily.' They were in the clear and were greeted by their fellow students with hearty applause.[6]

The technique of grave-robbery was standard among those well versed in the art. The approach was usually from the coffin side of the gravestone, but the thieves sometimes dug a tunnel under the stone from the rear, to obtain maximum cover and to leave the grave itself undisturbed. A hole was dug to expose about one-third of the coffin. A strong crowbar was inserted between the lid and body of the coffin, and powerful leverage either lifted or, more often, snapped off the lid. The corpse was then easily removed, the clothing stripped off and replaced in the coffin to avoid possible identification, and the grave filled in and tidied up.[7] Finally, the specimen was transported in a sack to its intended destination, a dissecting room. The whole procedure from start to finish usually occupied about an hour to an hour and a half.

It was along these lines that Pattison's medical students carried out their ghoulish activities, which were not lost on their university friends from non-medical faculties. The theological students, in their satirical periodical called *The Emmet*, published the first verse of a nasty little poem in 1824 entitled 'The Resurrectionist: A Tale Humbly Inscribed to the Editor of the Glasgow Chronicle':

'Twas a cold winter night, and dark was the clouds,
And the dead men lay quietly still in their shrouds;

The worms revelled sweetly their eyeholes among;
It was a rout night, and there was a great throng:
Some fed upon brains, others fed upon liver,
Had we e'er such a feast, all cried out, O! no never.[8]

Another example of graveyard poetry is found in a much wittier poem entitled *Mary's Ghost (A Pathetic Ballad)* by Thomas Hood, who, with some effective punning and a skilful combination of the humorous and the pathetic, conjures up a vivid picture of an unfortunate young woman's end. He refers to several of the leading medical figures and institutions of the day.[9]

'Twas in the middle of the night,
   To sleep young William tried,
When Mary's ghost came stealing in,
   And stood at his bedside.

O William dear! O William dear!
   My rest eternal ceases;
Alas! my everlasting peace
   Is broken into pieces.

I thought the last of all my cares
   Would end with my last minute;
But though I went to my long home,
   I didn't stay long in it.

The body-snatchers they have come,
   And made a snatch at me;
It's very hard them kind of men
   Won't let a body be!

You thought that I was buried deep,
   Quite decent like and chary,
But from her grave in Mary-bone
   They've come and boned your Mary.

The arm that used to take your arm
   Is took to Dr. Vyse;
And both my legs are gone to walk
   The hospital at Guy's.

I vow'd that you should have my hand,
    But fate gives us denial;
You'll find it there, at Doctor Bell's
    In spirits and a phial.

As for my feet, the little feet
    You used to call so pretty,
There's one, I know, in Bedford Row,
    The t'other's in the city.

I can't tell where my head is gone,
    But Doctor Carpue can.
As for my trunk, it's all pack'd up
    To go by Pickford's van.

I wish you'd go to Mr. P.
    And save me such a ride;
I don't half like the outside place
    They've took for my inside.

The cock it crows—I must be gone!
    My William, we must part!
But I'll be yours in death, altho'
    Sir Astley has my heart.

Don't go to weep upon my grave,
    And think that there I be;
They haven't left an atom there
    Of my anatomie.

Mrs McAllaster's Exhumation

Pattison's trial for body-snatching bears witness to the truth of
Hood's poem as a macabre comedy of errors. The court records,[10]
though confusing and illegible in places, allow reconstruction of the
crime and the ensuing events, leading to a dramatic climax before the
High Court of Justiciary in Edinburgh. At this time Pattison was a
rising lecturer at the College Street Medical School.

Mrs Janet McAllaster (or McAllister or McAlister), née McGregor, aged forty, died from consumption (pulmonary tuberculosis) on Wednesday, 8 December 1813. She was buried on Monday, 13 December in the newly constructed and fashionable middle area of the Ramshorn Churchyard, also known as St David's or the North-West Churchyard. The new burial ground, 'a churchyard within a churchyard', was constructed in 1788 and was facetiously termed 'Paradise' because it provided the last resting place of the rich and select. 'Lairs' (grave sites) in it were expensive, and it soon became the most fashionable burying ground of the city. Pierre Emile L'Angelier, the poisoning victim of the notorious Madeline Smith, was buried in the Ramshorn Churchyard in 1857, but was legally exhumed for pathological examination. This was the last case of body-raising in the long history of the churchyard. The location of Mrs McAllaster's lair is marked with a cross on the street plan.[11]

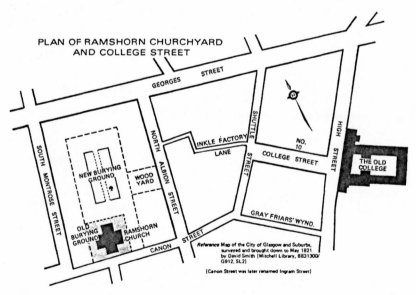

PLAN OF RAMSHORN CHURCHYARD
AND COLLEGE STREET

Reference Map of the City of Glasgow and Suburbs, surveyed and brought down to May 1821 by David Smith [Mitchell Library, 8831300/ G912, SL2]

[Canon Street was later renamed Ingram Street]

Following the funeral service, Henry Braid, the grave-digger, interred her in the presence of William Lindsay, the coffin-maker and undertaker, and John Robertson, the church officer. Later that evening a Janet Gunn was frightened in College Street by a crowd of students coming along Inkle Factory Lane ('I took a fright lest it should be these doctors'). The students were breathing as if in a hurry, but she was unable to see if they were carrying anything.

The following morning, Braid, who had gone to check the grave between 8.00 and 9.00 a.m., found that it had been desecrated. He informed Lindsay, who was still in bed, and together they examined the grave. Earth was scattered around 'as if a set of pigs had been rooting at it'. There were many shoe impressions and tracks, preserved by heavy frost, leading over the north-east dyke in the direction of College Street.

Braid and Lindsay next called upon Mrs McAllaster's two brothers, James and Dugald McGregor, since her husband, Walter McAllaster, was too broken-hearted to be of much help. With Robertson in attendance, the grave was dug to a depth of about six feet, and the coffin was lifted out and opened. The lid was found to be broken and the coffin empty except for Mrs McAllaster's shroud.

News of the theft soon spread and public reaction was swift. Some citizens blamed Professor Jeffray, and in their rage smashed all the windows in his house. At this time, Robertson met John Burns in nearby Argyle Street. Robertson strongly suspected that Pattison had the body in his rooms at the College Street Medical School (marked 'NO. 10' on the street plan), and proposed to Burns that, if this were so, it be returned quietly and privately to the graveyard. The proposal was reasonable, since Professor Jeffray had been known to give up stolen bodies on former occasions. Burns replied that Pattison had no particular need of the body, and he did not think he had it.

Burns, who himself had been prosecuted for grave-robbery some years earlier, agreed to send a message to Pattison by William Ruan, one of the apprentices at the College Street Medical School, warning him that a body raised from the High Churchyard was being urgently sought. But Burns's message contained an important error. The High Churchyard lies beside Glasgow Cathedral, at a distance of about half a mile from the Ramshorn Churchyard. Both churchyards had been robbed by resurrectionists on the night of Monday, 13 December, and Burns's message referred to the *wrong churchyard*. Ruan found Pattison at about 11.00 a.m. in one of the dissecting rooms at the School and informed him that if the body were given up, nothing more would be said. Pattison specifically asked Ruan for confirmation that the body in question really was from the High Churchyard and, on being assured on this point, stated that he did not have it.

John McLean and Robert Monro, two medical students, were present during the conversation between Ruan and Pattison. Monro, who usually did the dissections for Pattison's classes, was just about to open the abdomen of a woman's body. At the trial, Ruan was unable to given any indication of the age of the body, but he had noticed that the face was covered and the breast had become dark in colour.

Meanwhile, Dugald McGregor, although he wished if possible to keep the matter quiet, had arranged for a general search warrant for all anatomy rooms in the city. Accompanied by town officers, he and his brother went first to a George Monteath's anatomy rooms in the Gallowgate, where a decayed female subject was found, divided into two parts and lacking teeth. James McGregor was convinced that this was the body of his sister. However, James Alexander, a dentist who had made sets of teeth for Mrs McAllaster, stated emphatically that it was not her body.

Since Pattison was a prime suspect, porters had been set to watch his rooms at Walter McAllaster's expense from about 10.00 a.m. The search group, which now included Alexander and James Pirrie, the senior of the town officers, eventually arrived at 1.30 p.m. at Pattison's rooms at the College Street Medical School, armed with their search warrant. The group beat on the door violently with their walking sticks to no avail. They heard footsteps inside, approaching the door and then retreating. The officers were not empowered to break open the door since their search warrant gave no authority for forcible entry. During the delay, Pirrie dispatched sentries to guard the rear windows and doors of the school.

The group, finally admitted at 2.00 p.m. by Monro after nearly half an hour of knocking, entered with a rush. When asked for an explanation of the delay in opening the door, Monro muttered an inaudible remark while Pattison explained that the noise of some bellows McLean had been using had obscured the sound of the knocking. Pirrie with a straight face said the dead might have heard them.

First, Dugald McGregor was called into the lecture room by an officer to see an old woman with her throat laid open. He easily recognized that this was not his sister. He then proceeded to the dissecting room, where he saw Pattison, Monro, and McLean. Monro, who was wearing a striped frock gown with long sleeves, appeared extremely agitated and was sweating profusely. Dugald said that if the body of his sister were given up no further action would be taken. But Pattison replied that it was 'only private lecturers that lifted bodies', implying that he did not have it.

The searchers, who counted about six bodies in various stages of dismemberment, found a pan of human bones on a fire, and more bones in tubs filled with 'bloody coloured' water. Other pans contained entrails, a heart, a liver, kidneys, and lungs. Close by were earth-covered spades, a bludgeon, hats, ropes with loops and nooses, wet jackets and wet trousers, and three empty bags each large enough to contain a human body, all of which were labelled and removed for use as trial exhibits. One of the tubs on examination yielded three heads, the last of which was produced by Monro with extreme reluctance. Dugald immediately identified this third head as that of his sister, although it had been mutilated by the loss of the lower jaw, nose, upper lip, right ear, some teeth, both eyes, and the scalp. Inexplicable for anatomical purposes, such disfigurement is understandable as a means of concealing identity. Alexander, who had just trampled on an ear 'said to have been hers', seemed satisfied that the teeth were those of Mrs McAllaster, even with the mandible missing. Several witnesses stressed that the head in question was fresh and showed no signs of putrefaction.

At this point Dugald McGregor fainted. Pattison himself appears to have remained fairly calm but was much more disconcerted when the third head was produced. His hands trembled but he was not sweating. Monro was seen to remove his gown in a surreptitious manner. One of its pockets was later found to contain an underjaw wrapped in two handkerchiefs, one of which had his initials embroidered in a corner.

Word of what had happened in Pattison's rooms reached people outside. An angry mob gathered in the street, and military personnel had to be called out from the barracks in the Gallowgate to quell the disturbance. Pattison, whose arms were held firmly by Dugald McGregor and Alexander, and his two students were arrested and escorted under military guard to the Council Chambers according to the warrant. On the way, they were pelted with stones. All three were examined and signed official declarations, after which they were released on bail, Pattison in the amount of £100 and Monro and McLean £30 each.

Soon after his release, Pattison returned to College Street in company with Dugald McGregor and Dr Robert Watt, a physician and colleague at the School, to examine the various specimens thought to have derived from Mrs McAllaster. McGregor said he 'knew her as well as he did before she died'; Pattison denied that it was Mrs McAllaster's body.

When all except Pirrie had dispersed, Walter McFarlane, one of the porters who had been hired to watch the school since the morning, busied himself in conducting a thorough search through the anatomical bric-à-brac for any missing parts. In this he was successful, adding several items to the composite body that was gradually taking shape. His most significant find occurred at about six o'clock with the discovery of a female pelvis, which was to figure so prominently in Pattison's defence.

Fresh troubles confronted Pattison the next day, Wednesday, 15 December. He was arrested and jailed again when the procurator fiscal (public prosecutor) demanded on the basis of Tuesday's findings that his bail be increased to £200. This was quickly paid. Meanwhile activity at the School consisted of continued searches of all possible hidden locations and repeated attempts to fit the various anatomical pieces into one convincing whole. And it was on Wednesday that three medical men carried out the first of several examinations of the body. Watt, one of the three, thoroughly investigated the uterus and ovaries but made no mention of this in the report.

Thursday was occupied with further finds and fittings and with visits from the medical examiners. As each new development occurred, Alexander bustled in to have a look. Later in the day, the hands were taken by the officers to the Council Chambers to be shown to some friends and relatives of Mrs McAllaster, all of whom claimed to recognize characteristic marks and defects.

Also on Thursday, Dugald and James McGregor carried out their own detective work. First, they showed that some of their sister's shoes fitted the feet in the College Street rooms 'if clothed in stockings'. Next, they cut off a small lock of hair from the head in College Street and found that it appeared to be identical to some hair taken from Mrs McAllaster at the time of her death. Finally, Dugald studied the freckles on the reconstructed head and was more than ever convinced that it was his sister.

By Friday, 17 December, most of the work had been done. The officers and doctors performed a final fitting of the body, after which it was placed in a new coffin. The three doctors wrote their report. Only then were the officers dismissed who had been guarding the rooms since the Tuesday. The eventful week concluded the next day with the reburial of the body at two o'clock, 'in a handsome and expensive style'. That the body in fact derived from at least two individuals was to become manifest during Pattison's trial.

The Declarations and Indictment

After his arrest with his two students on Tuesday, 14 December, Pattison had been in serious danger of being lynched by the furious crowd. 'It was with considerable difficulty, and amidst horrid yells, that they were safely landed in the Tolbooth.'[12] Immediately after their arrival, Pattison was examined by Archibald Newbigging, one of the Glasgow magistrates, in the presence of two witnesses.

He answered a few routine questions about the lectures which he and Andrew Russel gave in the College Street Medical School, admitting that 'as a matter of necessity human bodies must be used to illustrate the anatomical lectures or at least part of them'. He also allowed that there were at the moment in his rooms the bones of some subjects which had been dissected. On being questioned further about the bodies of Mrs McAllaster and other females, he declared that 'having consulted with a man of business, he declined answering this question or any other question relative to the matters in the procurator fiscal's application'. Each of the three pages of the declaration was signed by Pattison and the magistrate.

McLean and Monro made and signed identical declarations. They stated that they attended the surgical and anatomical classes of Russel and Pattison, and that on occasion they assisted in dissections. Beyond that, they declined to answer any question whatsoever.

After these declarations had been completed, the three prisoners were released. Several weeks later, Pattison was recalled and re-examined by William Rodger, another of the Glasgow magistrates. In this second declaration he was questioned about a human body, or parts of one, subsequently taken from his dissecting rooms for reinterment. In answer, he declared 'that it is his determination not to answer any questions that may be put to him; that this does not proceed from any disrespect to the court, for which he entertains proper sentiments of respect; neither does it proceed from the declarants being guilty of any of the matters charged in the procurator fiscal's application, the declarant being entirely innocent, as he does now and has at all times averred; but it proceeds from his respect for his profession, and from a wish not to establish a precedent for answering upon these delicate subjects'.

The last person to be accused was Andrew Russel. He was examined on Wednesday, 29 December 1813 by William Dalglish, yet another of the Glasgow magistrates. It emerges from his declaration that Walter McAllaster had petitioned the magistrates to include Russel among the accused, from his 'being guilty, or accessory to the guilt of other persons, in the theft of the body of his wife from its grave in the North West Churchyard'. From the questions put to Russel, it seems likely that he was accused solely because of his partnership with Pattison, by which he shared in the profits of Pattison's lectures. Russel admitted nothing; like the others, and on the same grounds, he refused to answer any questions, one of which suggested that he had helped with the mutilation of Mrs McAllaster's body on Tuesday, 14 December.

The magistrates speedily reported these matters to the law officers of the crown in Edinburgh. The result was an indictment of Pattison, Russel, Monro, and McLean, at the instance of the Lord-

Advocate (the Solicitor-General for Scotland), to stand trial for the criminal abstraction of the body of Mrs McAllaster. The actual indictment is a long and intimidating document running to ten pages each of which is signed by Henry Home Drummond, Advocate-Depute (A.D.)[13] It demonstrates the leisurely legal style of those days:[14]

> *Granville Sharp Pattison*, Surgeon and Lecturer on Anatomy in Glasgow, *Andrew Russel*, Surgeon and Lecturer on Surgery there, *Robert Monro* and *John McLean*, both Students of Anatomy in Glasgow or lately Students of Anatomy there:
>
> *You are indicted and accused*, at the instance of Archibald Colquhoun of Killermont, His Majesty's Advocate, for His Majesty's Interest, *that albeit*, by the laws of this and of every other well governed realm, *the violating the Sepulchres of the Dead*, and the raising and carrying away dead bodies out of their graves, *more especially* when the said dead bodies are mutilated and disfigured for the purpose of concealment, is a crime of a heinous nature and severely punishable; *yet true it is and of verity* that you the said Granville Sharp Pattison, Andrew Russel, Robert Monro and John McLean are all and each or one or other of you, guilty of the said crime, aggravated as aforesaid, actors, or actor, or art and part [i.e., as principals or accessories]; *in so far as*, Janet McGregor, wife of Walter McAllaster, Merchant in Glasgow, having died on the eighth, or on some other day of the month of December, one thousand eight hundred and thirteen, and having been buried on the thirteenth or on some other day of the said month, in the North-West Churchyard, or burying ground, adjoining to the said church, in the city of Glasgow, you the said [accused] did, all and each, or one or other of you, on the said thirteenth day of December, or on one or other of the days of that month or of November immediately preceding or of January immediately following, in company with one or more persons your accomplices, to the Prosecutor unknown, proceed to the said Churchyard, or burying ground, and did then and there, wickedly and feloniously raise up and take out of the coffin and

The first page of Pattison's indictment to stand trial for 'violating the sepulchres of the dead'. Scottish Record Office JC 26/368. (Published with the approval of the Keeper of the Records of Scotland) grave the dead body of the said Janet McGregor, and did carry away the same to the [lecture and dissecting rooms] in College Street . . . ; and the said grave having been discovered to have been opened, and the body to have been amissing the next day thereafter, and a suspicion having arisen that the body was deposited in the said lecture and dissecting rooms, a warrant was applied for and obtained to search the same, and you the said Granville Sharp Pattison, Robert Monro and John McLean conscious of your guilt, being within the said lecture and dissecting rooms on the said next day thereafter, when the Officers of Justice arrived to search for the said body, did refuse them admission for the space of nearly half an hour though armed with a legal warrant, during which time you were occupied in mangling and disfiguring the said dead body in a horrid and disgusting manner for the purpose of concealment, in which state it was thereafter found and identified in the said lecture and dissecting rooms.

There follow details of the five declarations emitted by the accused; acknowledgment of a medical report on the body, dated Friday, 17 December 1813, by Drs Robert Cleghorn, Robert Watt, and William Anderson; and a list of the various exhibits to be used in the trial, which included two sets of artificial teeth made for Mrs McAllaster's upper jaw and all the paraphernalia found in the College Street rooms. The indictment concludes:

At least, the dead body of the said Janet McGregor was time and place foresaid wickedly and feloniously raised and taken out of the coffin and grave in which it had been buried, and was carried away, and was mutilated and disfigured for the purpose of concealment all as libelled, and you the said [accused] are all and each, or one or other of you guilty thereof, actors, or actor, or art and part: All which, or part thereof, being found proven by the verdict of an assize [jury] before the Lord Justice General, The Lord Justice Clerk, and Lords Commissioners of Justiciary, you the said [accused] ought to be punished with the pains of law to deter others from committing the like crimes in all time coming.

The indictment, which is undated, also contains a list of thirty-five witnesses for the crown, with a note of their various occupations.

Pattison, understandably, did not go out much in the six months between the crime and the trial, but he and Russel continued with their regular activities during this their first year of joint teaching at the College Street Medical School. They were honoured at a ceremony held six weeks before the trial. The *Glasgow Herald*[15] of 25 April 1814 reports: 'On Friday last, Mr. Russel, lecturer on surgery and physiology, and Mr. Pattison, lecturer on anatomy, were presented by their students with two elegant silver cups, having suitable inscriptions, as a testimony of respect for the ability with which these courses were conducted.'

Originally the trial was to have been held in Glasgow before the Circuit Court, but Pattison was opposed to this location, partly because he believed that it would be hard to find an impartial jury in his home town and partly because he felt that his reputation and practice would suffer. McAllaster, on the other hand, favoured Glasgow, and was even prepared to raise an action in his own name and at his own expense. Pattison was ultimately successful in having the trial transferred to the High Court of Justiciary in Edinburgh. After it was all over, McAllaster requested the Lord-Advocate to refund some expenses he had incurred in hiring officers and gathering information, and added in support of his request that he had been offered large sums of money to drop all further proceedings; he had refused the bribes. But the Lord-Advocate rejected his claim.

As the day of the trial approached, 'vast interest was manifested about it in Glasgow by all classes, from the highest to the lowest'.[16] All coaches from Glasgow to Edinburgh were crammed on the Friday and Saturday preceding the trial; and the Mail coach itself, on Sunday, had for a month before been fully reserved by the authorities. Great crowds gathered at the Black Bull and Tontine hotels to see the witnesses safely away. 'But the accused themselves did not dare to show face. . . . They travelled *incog.* to Edinburgh some days before the trial.'[17] On the day of the trial, hundreds tried to obtain admission, but many were turned away.

The Trial

The trial of Pattison and his associates was both sensational and of considerable legal significance. That five judges sat is an indication of the importance of the case; at that time the judicial strength of the High Court was only six. It is referred to in what is the most respected authority on the older Scottish criminal law, Hume's *Commentaries on the Law of Scotland Respecting Crimes*.[18] It made medico-legal history on two major counts.

At ten o'clock on the morning of Monday, 6 June 1814, the Lords of Justiciary (the Supreme Court of Scotland) took their seats in the old Justiciary Hall in Edinburgh. The presiding judge was Lord Justice-Clerk David Boyle, assisted by Lords Hermand, Meadowbank, Gillies, and Pitmilly. Counsel for the Crown (prosecution) were Alexander Maconochie, His Majesty's Solicitor-General for Scotland, Henry Home Drummond, Advocate-Depute, and Francis Jeffrey, Advocate. Counsel for the Defence were John Clerk, Advocate (later Lord Eldin), and Henry Cockburn, Advocate (later Lord Cockburn).[19] Each of the four accused was guarded by two men with drawn bayonets.

The trial began with the reading of the indictment to the accused in open court, after which each pleaded 'not guilty'. Cockburn then stated that they rested their defence upon a denial of the facts charged against them, and that all of them 'were so engaged and in such places at the time when this crime, if committed, must have been perpetrated, that it is therefore impossible it could have been committed by them'.

The judges then expressed their opinion that they found the indictment 'relevant to infer the pains of law' (i.e., legally valid if the facts alleged were proved to the satisfaction of the jury). The accused were allowed 'to prove all facts and circumstances that might tend to exculpate them or alleviate their guilt'. A jury was named consisting of fifteen Edinburgh men, whose occupations included architect, merchant, haberdasher, ironmonger, banker, clothier, grocer, bookseller, and wine merchant.

Before the full legal proceedings could begin, Clerk, expressing his regret at having to disappoint the 'laudable curiosity of such a numerous assembly', moved that the court be cleared and the trial proceed behind closed doors, on the ground that circumstances would be brought forward of a peculiarly gruesome and indelicate nature. In support of his motion, he referred to the act of parliament which required trials to proceed in public, except those involving licentious cases such as adultery, rape, and the like (Act of the Scottish Parliament, 1693, c. 27). He felt that the present trial qualified as such an exception.

The prosecutor made no objection to this proposal, leaving the matter to the judgment of the court. Their lordships ruled that the exception did not cover this case but related only to those cases in which the evidence is of 'such a kind as may vitiate the minds or endanger the morals of the hearers: this is not true of anatomical details'. There was thus 'nothing in the case that could warrant them to exclude the lieges, although their lordships individually expressed their earnest wish and desire that as little publicity as possible should be given to it by newspaper editors, or others, as from the delicate nature of it, the full publication would only tend to inflame the minds of the vulgar'.[20] They therefore ordered the trial to proceed openly in the usual form.

After this diversion, 'the Procurators [counsel] for the prosecution proceeded to adduce . . . witnesses in proof of the Indictment, who, being all lawfully sworn purged of malice and partial counsel, emitted their depositions viva voce in presence of the Court and Jury, without being reduced into writing'.[21] This lack of a trial record and their lordships' directive to suppress publicity resulted in a testy comment in the Glasgow newspapers: 'Our reporter was therefore told that he could not be allowed to write down notes of the evidence; which prevents us from laying the case fully before our readers.'[22]

The case for the prosecution was apparently quite clear, cogent, and overwhelming. The events of the week of 13 December 1813 were described and amplified by the prosecution witnesses. Much emphasis was placed on the identification of their late sister by Mrs McAllaster's two brothers. Jean Scott, Dugald McGregor's wife, gave convincing evidence regarding her late sister-in-law's hands. She particularly referred to sores on the left thumb and a bruised thumbnail that had turned black; she identified both these features on a left hand found in the College Street rooms, and was absolutely positive that the hand was that of Mrs McAllaster. A friend gave very much the same evidence, referring to the nail and flesh lesions and to a characteristic crookedness of the fingers.

The medical report written by Robert Cleghorn, Robert Watt, and William Anderson was not helpful to the accused. It contained a brief note (but no description) concerning a small female pelvis containing the uterus and ovaries. Cleghorn answered questions about an examination they had been asked to perform on a dead body. He had known Mrs McAllaster and 'was very much inclined to think it was her body' for three reasons: the size; the appearance of the skin; and the good fit of a set of false teeth. He thought that all the parts he had examined were from one body. Finally he alluded to the mangled state of the body and expressed the opinion that the mutilations were inconsistent with any reasonable scientific work in anatomy.

The medical impressions of John Gibson, a surgeon, were similar. He had known Mrs McAllaster and thought that a head he had been shown was hers. He had also been shown the hands, feet, and trunk of the body, and they all 'appeared to fit'. Finally, he too referred to the good fit of the false teeth.

The most impressive evidence for the prosecution was that of the dentist, Alexander, whose precedent-setting use of dental records is now a familiar forensic method. He commented, perhaps naïvely, that a set of artificial teeth that fits one mouth will not fit another. He then attempted to show that a set of teeth which he had recently made for Mrs McAllaster's upper jaw fitted the head found in the College Street rooms, even though some of the teeth had been forcibly removed since her death. (The prevailing painful practice was to file down or break off the natural teeth at the gum margins prior to constructing artifical teeth.) Moreover, one of the artificial teeth had a characteristic gold pivot, which seemed to fit perfectly the appropriate place in the jaw. He further claimed that when the denture was in position, it articulated with the mandible found in Monro's pocket. Finally, he confirmed the identity of the dentures by producing a mould of beeswax which he had prepared for Mrs McAllaster. His conclusion was firm and definite: the head he had examined was the head of Mrs McAllaster. The testimony of James Alexander was a landmark in forensic odontology, being the first recorded instance of a dentist testifying as an expert witness and of dental evidence providing a means of identifying a corpse.

During the examination of the prosecution witnesses, John Clerk, the senior defence counsel, asked relatively few questions in cross-examination. He did, however, refer rather mysteriously on several occasions to the fact that Mrs McAllaster had been the mother of eight children. There was no dissension on this point.

When the prosecution concluded its case, the future looked bleak for Pattison and his colleagues. But Clerk, who now led for the accused, showed himself worthy of his reputation as one of the most brilliant advocates of his day. He first proceeded to prove Pattison's alibi for the night of Monday, 13 December 1813. At this time Pattison was living with his mother at 2 Carlton Place, a ten-minute walk from the school, on the other side of, and overlooking, the River Clyde. From the evidence of Jean Frazer, a servant at 2 Carlton Place, of Laurence Black, a clerk in the Royal Bank, of Mary Copland, a fellow guest at a party which Pattison attended, and of Robert Neilson, a Glasgow surgeon, Pattison's movements and activities became clear.

5.00–6.00 p.m.   He gave a lecture in the College Street Medical School on the muscles of the back and neck.

6.00–7.00 p.m.   Dressed in pantaloons, he had dinner with his mother at 2 Carlton Place.

7.00–8.30 p.m.   He rested, and changed his clothes for a party. Jean Frazer stated that he did not leave the house during this time.

8.30 p.m.   He was seen to leave the house, dressed for a party.

8.30 p.m.–1.15 a.m.   He attended a ball given by a Mrs Bell in Carswell Court, off George Street. (This is about a mile from Carlton Place, and only about 200 yards from the College Street Medical School.) There were eighteen or twenty people at the ball. Mary Copland and Laurence Black both stated that Pattison was 'never out of the room'. He did not dance.

1.15–1.30 a.m. He walked with Mary Copland to her home in the Trongate and was then seen by Jean Frazer on his arrival back at 2 Carlton Place. His clothes were not soiled.

1.30–9.00 a.m. Jean Frazer stated that he was at home all night.

9.00 a.m. He had breakfast on the morning of Tuesday, 14 December.

Pattison's time was fully accounted for; he could not have taken part in Mrs McAllaster's exhumation. He was dressed in formal silk evening attire—not at all appropriate for grave-robbery. From Janet Gunn's evidence, it is probable that the crime was committed just before 10.30 p.m.; this is substantiated by the preservation of the grave-robbers' footmarks by a heavy frost which was known to have set in before midnight. Braid, the grave-digger, had given as his opinion that it would have taken the culprits at least an hour to remove Mrs McAllaster's body. Pattison's absence from the party for so long a time would scarcely have passed unobserved; the evidence regarding his unbroken attendance at the party was crucial.

Clerk's next task was to try to disprove the medical and dental evidence. His first witness was Pattison's friend and colleague, Dr Robert Watt, who was the lecturer on the theory and practice of medicine at the College Street Medical School. His evidence could hardly be considered as having been 'purged of malice and partial counsel'. He testified that on the Tuesday evening he went to the College Street rooms where he found about twenty people, 'a promiscuous crowd'. He walked in freely and nobody questioned his right to be there. He noted that several of those present were very active in trying to fit together a great number of anatomical parts derived from three or four bodies, all in varying stages of dissection or putrefaction. His overall impression was one of confusion. Watt then noticed Alexander, the dentist, who 'seemed particularly anxious in fitting artificial teeth'. Watt did not think they fitted well 'as some small stumps appeared to come in the way'.

Accompanied by Cleghorn and Anderson, Watt had returned the next day by warrant from the magistrates and had made an extremely significant observation about the pelvic contents of the body. In his opinion, the uterus was nonparous (having borne no children) and had never been in a state of impregnation, and the ovaries showed no trace whatever of corpora lutea or of scar tissue. (It was known, even then, that when pregnancy takes place, certain anatomical features called corpora lutea are always found in the ovaries; after about a month these become permanent scar tissue.) Watt concluded that the body was that of a seventeen-year-old virgo intacta. Like the use of dental evidence in the testimony of Alexander, the characterization of a cadaver by such means was of great significance in medico-legal history.

Watt testified that he and his colleagues returned to the College Street Medical School on Friday, 17 December to make their final report. On this visit, Alexander 'succeeded better than they did' in fitting the artificial teeth. On cross-examination, Watt stated that the general condition of the anatomy rooms was perfectly normal except for a little carelessness and want of cleanliness, and was no different from those of Professor Jeffray. He added that it was quite common to wear a gown while dissecting, as Monro had done, and for anatomists to put parts of a body in the pockets for future use. As for all the spades and bags, they were commonly used for the reinterment of dissected bodies legally obtained.

William Anderson corroborated Watt's testimony in all essential details. Most important was his confirmation of the findings about the uterus and ovaries. He asserted that there was nothing in the room inconsistent with the common state of a dissecting room, but admitted that he knew of no anatomical purpose for cutting off the tip of the nose and upper lip, or of violently removing teeth from a jaw.

The last two expert witnesses, James Scott and Robert Nasmyth, were dentists. Doubtless, defending counsel expected that their testimony would destroy the credibility of Alexander. Scott testified that, when he arrived accompanied by Pattison and two other doctors, the town officers tried to bar their entrance. 'One of them pulled out a pistol when we had entered and held it at my breast. This made us stay shorter.' But Scott saw enough to be able to deny resolutely that Alexander had succeeded in fitting the false teeth to the palate. Nasmyth, in turn, refuted Alexander's evidence that the upper artificial teeth articulated with the natural teeth in the mandible produced, and added that the denture fitted equally well four different heads.

All that remained were the final charges to the jury. The Solicitor-General, 'in a most ingenious speech', admitted that there was 'no evidence tending to inculpate McLean; but contended that the other three pannels [accused] were guilty of the offence charged, various parts of Mrs. McAllaster's body having been found in the anatomical school of Mr. Pattison, with which the other two pannels were, from the circumstantial evidence in the case, so connected as to render them all guilty, actors, or art and part, in the eye of the Law'.[23]

For the defence, Cockburn contended that no part of the evidence fixed upon Russel or Monro the slightest suspicion of their being concerned in the only offence charged, that of raising and carrying off the body in question.

Clerk, as counsel for Pattison, first addressed the jury on the evidence supporting the alibi. By a methodical review, he proved that Pattison could not have taken part in the illegal exhumation. Next, he drew the attention of the jury to the wording of the indictment, in which the gravamen was the *actual body* of Mrs McAllaster, and reminded them that all of the witnesses had stated that she had been the mother of eight children.

Clerk never denied that a grave or graves had been violated. What he did state most emphatically was that 'the body produced was NOT the *body* of Mrs. McAllaster at all; and, therefore, that the indictment must fall to the ground'.[24] In support of this he gave special prominence to the evidence presented by a number of the most eminent medical men in Glasgow. They solemnly swore that, after the most careful analysis, they were perfectly satisfied that the body produced was not the body of a married woman who had borne eight children but was that of a virgin. 'True,' they said, 'the *teeth* were identified, and the *fingers* were identified; but the Court and Jury must attend to the fact, that the prisoners under the indictment were not accused of abstracting either the fingers or the teeth, but the actual *body* of Mrs. McAllaster; and could they believe that this body, . . . sworn to and demonstrated by *clinical* evidence as that of a virgin, was really that of a married woman?'[25]

Following this dramatic defence, the Lord Justice-Clerk summed up the evidence in his charge to the jury:[26]

> This is a trial of grave and serious importance, involving the interests of the community, and the peace and comfort of individuals; you, the jury, will therefore give it particular attention. The law of Scotland distinctly lays down the violating of the sepulchres of the dead as a heinous offence. The prejudices and feelings of individuals are most naturally and justly revolted at the perpetration of such a crime. You will, at the same time, guard with due care your minds from any bias arising from feelings on this subject. The indictment, in the major proposition, charges the pannels with having raised up, and taken out of the grave, the dead body of Janet McGregor [McAllaster]; if this be proved, you must give a verdict for the prosecutor; but if not, whatever circumstances may have been disclosed, you can have no hesitation in returning a verdict for the prisoners. I certainly expect to find the charge against Russel fallen from, since there is not a shadow of evidence against him, while against McLean there is a circumstance proved of some weight, but of which nothing can now be said, since the prosecutor has withdrawn his charge against him. You will therefore consider whether you should not find both Russel and McLean not guilty. It remains that a few observations be offered on the proof as to the other two. That the body was raised cannot be doubted, and it is for you to consider if there be not very satisfactory proof of the identity of the body found in the dissecting room with that raised from the grave. The evidence arising from the case of artificial teeth seems exceedingly strong; and that from the mark upon the hand appears no less satisfactory. It is vain to talk of uncertainty and error on the part of the witnesses on this subject; people can recognize such marks with the most infallible correctness; there is satisfactory evidence [in] a thorough conviction that persons who have been familiar with the deceased can derive from such appearances that no general arguments can remove

or shake, and which seems to leave no doubt as to the identity of the body. The circumstances too of the case—refusal of admission [to the College Street rooms]; confusion of the accused; the state in which the body was found—cannot fail to make an impression. Notwithstanding, therefore, the opinions adduced in exculpation, you will consider if there be not satisfying evidence of the identity. The next, and by far the most material part of the case, is the evidence adduced to prove that this body was actually raised up and carried away by the prisoners. It is for you, the jury, to weigh attentively all the circumstances of this part of the evidence, and likewise the evidence adduced to prove an *alibi*. I leave it to you to consider if, upon the whole, the proof be such as to convince you that the pannels, or any of them, engaged or participated in raising the body, which constitutes the crime. If you think there is no proof, and that the body was not identified, you will bring in a verdict of *not guilty*. If you think the body was identified, but that the proof was not sufficient to convict the pannels of participating in raising it, you will bring in a verdict of *not proven*.[27]

By this time, the trial had proceeded for sixteen hours, with only brief breaks. (It was customary, at this period in Scots law, for an entire trial, with the exception of the verdict, to be completed in one session.) It was 2.00 a.m. on Tuesday, 7 June 1814, when the jury was excused with these words: 'The Lord Justice-Clerk and Lords Commissioners of Justiciary *ordain* the assize instantly to inclose in this place, and to return their verdict in the same place this day at two o'clock afternoon, *Continue* the diet against the pannels till that time, *ordain* the whole fifteen assizers [members of the jury] and all concerned then to attend, each under the pains of law.' Other formalities took up another hour. The prisoners had to be re-admitted to bail before they could leave for the twelve hours between the end of the trial and the announcement of the verdict: this was set at £200 for Pattison and Russel, £60 for Monro and McLean. Finally the court adjourned 'past three o'clock on Tuesday morning'.

At two o'clock on the afternoon of the same day, the court reconvened. In the interim, the jury had elected a 'chancellor' (foreman) and a clerk, and had considered the evidence presented by the prosecutor and by the defence. They were thus able to present to the Lord Justice-Clerk a signed verdict: 'We all in one voice find the said Andrew Russel and the said John McLean *not guilty*, and find the libel *not proven* against the other two pannels Granville Sharp Pattison and Robert Monro.'

Lord Justice-Clerk David Boyle then delivered a homily to the prisoners:

> Granville Sharp Pattison, Andrew Russel, Robert Monro and John McLean, before reading to you the judgment of the court, I think it proper to address to you a few observations on the circumstances that have been disclosed in the course of your trial. The crime with which you have been charged is justly exposed to odium and punishment. It is undoubtedly necessary that human bodies be dissected. The purpose of an honourable and useful science renders this indispensable, but it must not be obtained by offending the feelings of individuals, and disturbing the repose of the tomb. The interests of the public, the feelings of humanity, call for punishment on the rude hand that molests the security of the grave, and that, by thus invading the silent repositories of the dead, violates the peace and disturbs the tranquillity of the public. Of this crime, two of you have been, by the unanimous verdict of a most respectable jury, found not guilty. I therefore congratulate you, Andrew Russel and John McLean, on your acquittal of all participation in the crime charged, and trust that the present trial will have the salutary effect of teaching you caution and care in the future pursuits of your profession.
>
> You, Granville Sharp Pattison and Robert Monro, are likewise acquitted by a unanimous verdict, finding the charge against you not proven. You will not, however, suppose the prosecution at the instance of His Majesty's Advocate improper or unjust. You were all of you most properly included in the indictment. You are now acquitted by the verdict of a

respectable jury, and from the present trial I trust you will learn caution and circumspection in your future conduct. You will not consider yourselves entitled to violate the graves of the dead, even for the purposes of science. You, Mr. Pattison, will be particularly careful and circumspect in the management and conduct of that profession which you hold, with much advantage, I have no doubt, to the public, and honour to yourself. The order and state of your class you will arrange so as to avoid all suspicion, and be careful not to place yourself in a similar situation again; for it must not be concealed, that circumstances were disclosed in evidence, which clearly prove that you had recourse to practices by no means necessary for any regular or anatomical purposes. In procuring subjects for dissection, you will adopt such measures as can give no offence to individuals, and do no discredit to yourself; for, from the doctrine laid down by your counsel, that the only remedy for individuals is to watch the graves of their friends, I strongly dissent. Such necessity is not imposed on private individuals; the public are bound to afford them protection and redress. In a churchyard, surrounded by houses, squares and streets, as we know to be the case with the North-west churchyard in Glasgow, how dreadful might the consequences be if a watch placed there with loaded guns were obliged to fire during the night. The public safety would never permit such measures. The graves of the dead must not, therefore, be invaded, and subjects for dissection must be otherwise procured. Permit me, in conclusion, to say, and I believe I have the concurrence of all my brethren when I say, that I hope and trust that no injury whatever to your character or success in your honourable profession will arise from this trial; but that, on your return to your usual employment, you will be respected and useful.[28]

The pannels were completely absolved and dismissed from the bar. The invidious verdict of 'not proven', which still survives in Scottish criminal law, has been sardonically translated as 'not guilty—but don't do it again'. It means 'not proven guilty (but not necessarily innocent)', leaving a permanent aura of doubt hanging over the accused.

From the Lord Justice-Clerk's instructions to the jury, it is clear that Pattison would have been found guilty only if it had been shown that he had participated in the actual grave-robbery. That he was innocent of any personal involvement in the crime was proved by his alibi, although even that may be slightly suspect, based as it was on the evidence of a maid and a female friend, especially as he was known to be young, attractive and very persuasive.

A verdict of 'not proven' was suggested by the Lord Justice-Clerk only if the jury considered the body in Pattison's rooms to have been that of Mrs McAllaster. John Clerk's dramatic use of medical evidence concerning the uterus and ovaries was directed at generating sufficient confusion about the identity of the body to justify an absolute acquittal of 'not guilty'. The jury apparently were not impressed by Clerk's ingenious sophistry, and considered the head and fingers to constitute satisfactory proof of the presence of Mrs McAllaster's body in Pattison's dissecting rooms.

It is ironic that the whole disagreeable chain of events might have been avoided, had not Pattison been misled by Burns and Ruan about the churchyard from which the body had been abstracted. Under the circumstances, he might well have realized that he indeed had the body that was causing all the trouble and have given it up quietly for re-interment. His failure to bow to expediency led inevitably to the search of his rooms, McFarlane's production of the 'wrong' pelvis (by accident or design), the identification of the head and hands, Watt's suspiciously delayed report on the uterus and ovaries, and eventually the trial itself.

Pattison's acquittal ended this unfortunate episode. It is interesting to consider what sentence he might have received had the verdict been 'guilty'. The laws of the time were harsh. On the same page of the *Scots Magazine* as the account of his trial are recorded sentences imposed in two other, arguably less serious, cases. Bertha Hamilton, convicted of falsehood, fraud, and wilful imposition, was sentenced to transportation for seven years. James McDougall, for uttering forged bank notes, was sentenced to be executed.

Violation of sepulchres was and is a crime at common law in Scotland, i.e., it was not made criminal by any statute but has been treated as punishable from the earliest times. Accordingly, no particular punishment has ever been prescribed for it, but it has always been punished by the courts under their common law powers. In his notes to his discussion of the crime of violating sepulchres,[29] Hume cites the case of Pattison and his colleagues, and refers to various sentences for the crime, including imprisonment with or without hard labour, fines, banishment from Scotland, and occasionally transportation beyond the seas. (Less than a year after the Pattison verdict, four Scottish medical students were convicted of raising the body of a woman from her grave and sentenced to fourteen days' imprisonment and a fine of £100.) It seems likely that, had Pattison been convicted and the Court disposed to treat him like others, the sentence would have been imprisonment, possibly with a requirement of finding caution for good behaviour, or possibly a substantial fine. Only if the offence had been regarded as specially heinous, as by reason of the accused's professional standing, would transportation have been likely. The death penalty does not seem ever to have been inflicted for this crime.[30]

The trial seems to have had remarkably little effect on Pattison himself, or on his professional career and position in society. Outwardly, at least, his attitude to the whole affair was cavalier; even fourteen years after the trial he referred to it in an off-hand manner. When asked, at a Select Committee Meeting of the House of Commons, if the police and magistrates were severe in punishing illegal exhumations, he replied, 'they behaved with the greatest severity; in my own individual case, the first year I taught, there was a body disinterred, and there was a skull without teeth found in my dissecting rooms; and because this person had had no teeth, I was dragged away by the police, carried through the populace, pelted with stones; I was then indicted, and tried like a common criminal in Edinburgh, a man sitting on each side of me with a drawn bayonet.' On being questioned about the outcome of the trial, he replied, 'An acquittal, which cost me £520 sterling.'[31]

The anguish, suffering and despair of Mrs McAllaster's relatives and friends should not be minimized. Clearly, all or part of her body had been stolen and carried to Pattison's dissecting rooms, there to be discovered in a scene of execrable horror amidst other human remains in varying stages of dissection or decay. Pattison, who was undoubtedly aware of the proposed exhumation, may well have initiated and planned the crime. But he was accused only of participating in the robbery, not of having her body in his rooms, and the jury gave the only verdict open to them. No miscarriage of justice is apparent in this important trial.

III
Professional Misconduct
and Personal Indiscretions
1814–1819

IT MIGHT BE SUPPOSED that Pattison's acquittal would have led to a period of relative tranquillity. At the time of the trial, he was only a twenty-three-year-old lecturer at the College Street Medical School, and his career as an anatomist and surgeon would depend on his future clinical and academic achievements. Progress in such work requires a peaceful atmosphere conducive to study and research. But this was largely denied him during his remaining five years in Glasgow. Indeed, during this period he was forced to defend not only his professional competence against a charge of malpractice but also his personal reputation against a charge of adultery.

He was also short of funds. He had apparently inherited little or nothing on his father's death in December 1807. He later wrote that his 'character' was the only legacy he had received from his father, and his only dependence. The trial had been expensive: on 11 August 1814, two months after the trial, Messrs Grahame and Mitchell, Glasgow solicitors, wrote to Pattison and Andrew Russel outlining the legal services they had provided during the trial and requesting 'that Messrs Pattison and Russel either settle these accounts in cash in course of tomorrow forenoon, or hand them a bill at a discountable date for the amount'.[1] Russel must have replied to this demand in a high-handed manner if we are to judge from the tone of

Mitchell's reply to him, dated 16 August: 'I am at a loss to account for the surprise you have thought fit to express at the demand made on you and Mr Pattison for payment of the debt owing by you and him jointly to Mr Grahame and me. That you and Mr Pattison jointly consulted me and employed me . . . we shall be at no loss to establish. We shall therefore insist for payment from you jointly and severally.' This letter must have had the desired effect because there is no record of further correspondence.

The Glasgow Medical Society

The pettiness of this financial bickering notwithstanding, Pattison and five other medical men met on Thursday, 27 October 1814, and formally inaugurated the Glasgow Medical and Surgical Society. His colleagues, two other surgeons and three physicians, were Dr Robert Watt (the first President), Dr Robert Graham, Dr John Robertson, Mr John Young, and Mr George Macleod. Watt, a fellow lecturer with Pattison at the College Street Medical School, had been an effective witness for the defence at his trial. Robertson, another teacher at the College Street Medical School, curiously, shared the same name as the Ramshorn church officer at the time of Mrs McAllaster's exhumation. Graham, who served as Pattison's physician, was later to be a member of a committee called to examine a charge against Pattison of unprofessional conduct at the Glasgow Royal Infirmary.

One month after its inauguration, the society was renamed the Glasgow Medical Society.[2] No reason has been found for this change; there is certainly no hint of strife among the physicians and surgeons. It is likely that then, as now, the term 'medicine' was recognized as proper to all aspects of the healing arts: a resolution at the inaugural meeting proposed 'that the society be exclusively for the prosecution of "medical science"'.

The rules of the society were severe and intransigent. Annual dues were one guinea, and members were fined one shilling for each meeting missed. Meetings were held on the first and third Tuesdays of each month, when scientific papers were read by its members in

'Numpskulls and Rumpskulls.' (Mitchell Library, Glasgow)
The members of the Glasgow Medical Society were undoubtedly sincere
and dedicated, but the public at times viewed their activities in a somewhat
suspicious light, as this cartoon from the *Northern Looking Glass* shows.

strict rotation. The subject of each was announced about two
months in advance, and the completed text was required to be
submitted to the society at the meeting immediately preceding that
at which the paper was formally read. Fines were imposed for any
alterations to a paper following its submission.

It must be remembered that knowledge of all the basic sciences
was then at a relatively primitive stage. Organic chemistry was in its
infancy, bearing little relationship to the sophisticated subject it is
today. All organic compounds, for example, were thought to be
derived only from living organisms which possessed a mysterious
'vital force'. Physics, in particular the concept of electricity, was
equally rudimentary: Faraday's basic work in this field came nearly
twenty years after the inauguration of the society. Fundamental
knowledge and concepts of physiology, bacteriology, and pathology were as yet undeveloped. One must feel sympathy for
physicians and surgeons of that time in their struggle to understand
and master the baffling disease phenomena with which they were

daily confronted. Equally deserving of sympathy, of course, were the patients, who were subjected to violent purging, blood-letting, the liberal use of various poisonous inorganic salts, and particularly the appalling agony of pre-anaesthetic surgery with its ever-present danger of iatrogenic infection.

It is in this light that the activities of the Glasgow Medical Society should be viewed. The founding members were obviously enthusiastic, and the society provided a valuable forum for the dissemination of information and the encouragement of free discussion.

The full texts of the three papers Pattison delivered to the Society have not survived, but the minute book contains a record of their titles: at the tenth meeting, 6 March 1815: 'Appearance of the Brain in Cases of Epilepsy'; the thirty-second meeting, 19 March 1816: 'Case of Affection of the Teeth with Observations'; the fifty-sixth meeting, 18 November 1817: 'Some Observations on Abdominal Operations'. Pattison's professional interests were clearly extremely catholic; it would be a courageous doctor who today would select for presentation to his peers topics in the apparently unrelated fields of neurological pathology, dentistry, and abdominal surgery!

Even as he was helping to found the Society, Pattison was confronted with financial, professional, and medical problems of his own. In November 'Pattison . . . in spite of most potent purgation, has complained of most excruciating headache, so that it is now thought prudent to take some blood from his jugular'.[3] This is the first evidence of the ill health which recurred intermittently throughout his life.

Furthermore, Pattison's relationship with Andrew Russel at the College Street Medical School had been deteriorating in the months following the trial. Their partnership finally dissolved in the autumn of 1814, and Russel emigrated to the United States, where he died in 1861. Before leaving he sold his share of the Burns museum to Pattison, who then became its sole owner. Armed with this outstanding teaching aid, Pattison became the main lecturer at the School. An advertisement in the *Glasgow Herald* of 7 October 1814 launched him in his first solo teaching position.

ANATOMY, PHYSIOLOGY AND SURGERY

On Tuesday, the 1st of November, at 5 o'clock in the afternoon, Mr Granville S. Pattison will commence a course of lectures on Anatomy, Physiology and the Principles of Operations of Surgery, in the Class Room, College Street.

The surgical lectures will begin on Monday, the 5th of December, and will be delivered on the Monday, Wednesday and Friday evenings, from 8 till 9 o'clock.

The valuable Museum, formerly belonging to the late Mr Allan Burns, and lately to Messrs Russel and Pattison, has been purchased by Mr Pattison, and will be used in illustration of the above lectures.

COLLEGE STREET   October 6, 1814

Glasgow Royal Infirmary

Pattison spent the following eighteen months building his reputation at the College Street Medical School as an anatomist, surgeon, and teacher. Part of this time was spent at the poorhouse, euphemistically known as the 'Town's Hospital', where he served as surgeon under John Burns's supervision. His role was as a substitute for the regular surgeon, and he was apparently neglectful of his duties. His resignation in the spring of 1816 was greeted with some relief by the Town's Hospital Committee: 'As he has resigned his charge, your Committee will abstain from alluding to what is past; but they consider it imperative on them to recommend, that on no account whatever in future, ought the surgical duties to be regularly performed by a deputy.'[4]

His resignation was occasioned by his appointment to the salaried position of junior first-year surgeon at the Glasgow Royal Infirmary. At this time there were three hospitals in Glasgow: the Lock Hospital, the newly opened Asylum for the Insane (later Gartnavel), and the Royal Infirmary. Founded in 1793, the infirmary was an elegant Adam building on the site of the former Bishop's Palace; it was later demolished to give place to the present-day mammoth

tribute to Victorian stonework. All activities were controlled by a
board of directors, the members of which served as managers of the
infirmary (the terms 'director' and 'manager' are synonymous in
this context). Their extremely strict regulations had a direct bearing
on the next dramatic crisis in Pattison's life.

Rules for patients were of Draconian severity.[5] 'If any patient
shall break these regulations, he or she shall be instantly dismissed
disgracefully, and can never again be admitted into the infirmary.'
Any person bringing food or liquor to a patient would be prevented
from visiting the infirmary 'for ever afterwards'. Moreover, any
patient who might give away or exchange any food furnished by the
infirmary would be 'expelled with infamy, and never admitted
again, however great the distress, or however strong the recommendation may be'. All patients were required to remain silent
when the physician or surgeon was visiting the ward but at the same
time to 'conceal no disease, and no circumstance relating to it'.
Those patients capable of doing so were to keep in order their own
beds and clothes closets. Visiting hours were restricted to two hours
in the morning; any patients 'harbouring visitors at other hours shall
be instantly dismissed'. And patients were warned not to disturb the
ward by talking loudly, quarrelling, swearing, 'nor even by smoking
tobacco'.

The house staff consisted of a porter, students, dressers, clerks,
nurses, and physicians and surgeons. The porter had to clean, open
the gates, accompany visitors, carry coals, trim lamps, ring bells,
remove bodies from the 'dead room', deliver letters and notes, and
assist all levels of the medical staff. In addition to all this, he was
expected to shave patients, administer injections, and take charge of
the patients' baths. And he was expressly forbidden to accept any
gratuity from patients or strangers, but was to 'gratify their desire
without fee or reward'.

Regulations for students were chiefly concerned with their
behaviour in the wards. 'Students are to behave with decency and
propriety, keeping their hats off, at all times avoiding doing anything that may disturb the physicians, surgeons, clerks or patients.'
And they were firmly forbidden to 'tease the patients with unnecessary questions' or to offer any advice or opinion to them concerning

their diseases. On these conditions they had the privilege of accompanying the physicians and surgeons in their daily duties, including operations and autopsies.

Dressers were appointed from among students with at least three months of hospital experience. They were required to attend every day and were 'on no account whatever to be absent without leave from the surgeon'. Each dresser had to have an apron with pockets and sleeves, a pan for clean dressings, and a small box for the soiled, discarded dressings; these last were to be carried off by the nurses on his order. Finally, he was required to serve as an assistant at all operations and autopsies.

The physicians and surgeons had clerks, senior students who were in charge of records and who performed 'all smaller operations' such as bleeding and cupping. The surgeon's clerk was in charge of the surgical instruments and was required to have an assortment of bandages in readiness at all times. He superintended and instructed the dressers, and was 'to be careful that there be no waste of caddis (woollen yarn), strap, or other materials', and he performed parts of the autopsies, which were 'to be conducted with the greatest decency, and the body left in a proper state for interment'. Operations and autopsies were to be listed by him in the students' waiting room one day in advance.

The nurses' role was menial. 'The nurses shall remove from their respective wards all dust and nastiness every morning before ten, and shall at all times keep them neat and clean, and be especially careful in airing them by keeping the upper sashes of the windows down.' In addition, they were required to make beds, dispense medicines, distribute meals and drinks to the patients, and report uncommon symptoms to the attending physician or surgeon. They were to be 'instantly dismissed with forfeiture of wages' if they brought into the infirmary any spirits or improper articles of diet; if they demanded or accepted any money, fee, or reward; or if they borrowed any money or clothes from anyone.

The general behaviour of the physicians and surgeons was of course much less rigorously controlled. They attended the infirmary daily at 1.00 p.m. to examine patients, write reports, and organize all surgical procedures. Operations had to be performed publicly in the presence of the students, 'excepting such as the attending surgeon may consider injurious to female delicacy'. Consultations were of vital importance to surgical patients since the decision to operate led inevitably to excruciating pain and the prospect of death through technical errors or infection. Some surgeons tried to avoid consultations either from a fear of exposing their ignorance or from a desire not to divulge any special techniques to potential competitors. It was in the actual conduct of operations that the strictest regulations were laid down. One of these expressly stipulated that no operation was to be performed unless approved at a consultation, regularly summoned.

The Hugh Miller Quarrel

It was in this setting and with these constraints that Pattison found himself in 1816. True to form, he was soon in trouble again.[6]

An intense antagonism had developed between himself and Hugh Miller, one of two senior staff surgeons and a director of the infirmary, who, Pattison later claimed, was determined to bring Pattison before the managers for unprofessional conduct. An opportunity to do so occurred on Wednesday, 11 December, when Pattison performed two amputations which resulted in the deaths of both patients. A few days later, Miller openly accused Pattison of incompetence. There followed a number of meetings of medical committees and of managers, culminating in Pattison's demand that a formal meeting of the managers be called to examine his conduct. At this meeting, on 26 December 1816, Pattison and Miller had a violent quarrel. Shortly afterward, Pattison challenged Miller to a duel, which was declined, whereupon, in the coffee room of the Tontine Hotel, he posted Miller for cowardice.

The board appointed a special committee consisting of the regular medical committee of the infirmary and three outside surgeons: John Burns (by now professor of surgery at the University of Glasgow), John Towers, and William Couper. They were instructed to 'enquire and report into Mr Pattison's matter'. Pattison was required to defend his conduct by preparing a statement of exculpation to be presented to the board's clerk prior to the enquiry. Miller, as a committee member, would serve as both accuser and judge.

The eight-man committee of enquiry met on Thursday, 2 January 1817, chaired by Dr Richard Millar, the senior physician at the infirmary. Millar informed the committee that all three of the outside surgeons had declined to attend. This gave Pattison the chance, in the initial skirmishing, to protest that the members present were not competent to proceed with the enquiry since the committee was incomplete as constituted by the board. He therefore refused to bring forward any exculpatory evidence whatsoever.

However, Pattison had failed to comply with the board's order to prepare and submit a statement prior to the enquiry. He explained this by stating that his amanuensis had been delayed in preparing a fair copy because of the illegibility of his rough draft. But the board's clerk testified that, when he had met Pattison a few days earlier, the latter had told him that he had been advised by a lawyer not to submit anything. Considerable arguing ensued but Pattison finally capitulated and voluntarily delivered his memorial to the committee. Hugh Miller remarked that, as he had not been able to study the memorial, he felt free to disregard it.

The committee agreed to ignore Pattison's protests and to allow the hearing to proceed. But by now so much time had been wasted that the meeting adjourned. The members reconvened at six o'clock the same evening at the Tontine Hotel, where they remained, 'examining witnesses and squabbling among themselves', until after five o'clock the next morning.

First to be considered was the case of John Young. Forty hours before his admission, he had fallen down a coal pit and severely fractured his right leg and knee. A consultation was called, attended by the physicians Robert Graham and Richard Millar, and the surgeons Hugh Miller, Harry Rainy, and Pattison. It was unanimously agreed that Young's life could be saved only by amputation, but there was a difference of opinion as to how high up the leg this should be performed. Miller, Millar, and Graham elected for a low femoral amputation, whereas Pattison and Rainy, believing that the femoral neck had been severely injured, favoured the more dangerous disarticulation at the hip. The majority view was thus for low amputation; disarticulation was to be performed only if femoral neck injury rendered it indispensable.

Next came the key issue of the enquiry. Pattison believed that he was given permission to tie the femoral artery *at the groin*, thus cutting off the blood supply to the entire leg and effectively ruling out the possibility of low amputation. But his accusers stated emphatically that permission for this procedure was conditional; it was to be done only if the low amputation had been attempted, with inadequate results.

After the consultation, Pattison tied the femoral artery just below the inguinal ligament and, only one and a half inches below this, made a circular perpendicular incision right to the bone, which he then sawed. But the nature and location of this incision made wound closure virtually impossible. Having tried unsuccessfully to cover the femoral stump with soft tissue, he asked his colleagues for permission to proceed immediately with the disarticulation. They all assented and this was done, thereby affording Pattison the satisfaction of having performed the operation he had originally preferred. Subsequent examination of the bone indicated no injury at the femoral neck. John Young died a few days after the operation.

Pattison's accusers made, directly or by implication, the following charges: (1) that he did not operate according to the majority opinion at the consultation; (2) that he should not have tied the femoral artery at the groin without first having performed the standard low amputation; and (3) that, in the amputation, he deliberately operated too high up and by the wrong technique to allow proper coverage of the bone by soft tissue.

The second case was that of Jean Gowdie, who had met with a horrible accident in a Glasgow textile factory. A plashing (trimming) machine with toothed wheels had caught her by the apron and pulled her in. It had torn off large areas of skin and fat from her right thigh, and had then caught her by the breast and had torn open the armpit, exposing nerves and arteries, opening the main veins, and tearing several large muscles.

She was first seen by a Dr George Watson, who noted the terrible wounds of her hip and axilla, with their huge flaps of skin and underlying tissues. She mentioned to him that her ribs might be broken and complained of feeling extremely cold and weak, but she was able to stand and to move her wounded arm and thigh. Her pulse was very weak, at a rate of forty per minute. Watson gave her some spirits and then sutured the worst of the thigh flaps back to its original position. She was transported to the infirmary in the evening.

Pattison was there to admit her, having just completed the operation on John Young. He at once removed Watson's sutures and cut off the entire thigh flap. He then tied some of the veins in the axilla and found that the clavicle was dislocated at the shoulder. At this point he sent out cards calling for a consultation, as required by the infirmary, marking the cards 'a case of dreadful laceration'.

Because it was late evening, some delay occurred before the consultants arrived. Eventually, a small group, including Pattison and Miller, held a consultation. Miller saw the huge flap of skin from the thigh but was not informed that it had been previously sutured in position by Watson and subsequently excised by Pattison. He did not examine the thigh wound, nor did he examine the patient's chest, though he had been told that her ribs might be broken. He did examine the axillary wound. His opinion was that 'the patient should be allowed to die in peace'. He heard no dissenting voice from the others present and concluded that all agreed. He went home after Pattison had declined his offer to remain in the infirmary.

After Miller left the infirmary a second consultation was held, which included Pattison and three other newly-arrived medical men. Their unanimous verdict was that, as soon as the patient's condition stabilized, Pattison should remove the arm at the shoulder, this being the patient's only hope for survival. Miller's contrary opinion was reported, considered, and rejected. Graham, who arrived later, agreed with the decision to operate. He was not, however, informed of the earlier consultation.

At one o'clock in the morning, Pattison and Rainy found that the patient's pulse was stronger and more rapid, and that she was now free of tremor. Pattison removed her arm by disarticulation at the shoulder, and then covered the exposed surface with a flap of healthy soft tissue. On his way to the operating room, he had remarked to Rainy that he 'never went to perform an operation with more painful feelings'. The patient was alive and feeling better the following afternoon; she died later, apparently from pulmonary complications.

At the enquiry, Miller complained that, at the first consultation, he was not informed of any proposal to amputate the arm. He learned with astonishment the following morning that the operation had been performed during the night. He had not been called to a second consultation, nor had any of the infirmary's senior physicians and surgeons been notified of the proposed operation.

Graham came to Pattison's defence. He asked Miller why he had not examined the chest after being informed that the patient's ribs might be fractured. Miller replied that George Watson had told him that the woman herself had stated her ribs might be broken, and that, from the nature of the injury, he had no doubt that she was right. He added that he did not consider it necessary or proper to put her to more pain by examining the seat of her ribs.

The charges made or implied against Pattison in this second case were: (1) that he had performed a dangerous and painful operation without authorization from a properly constituted consultation, as required by the infirmary's regulations; (2) that he had not called a consultation to determine if the patient were in a state to bear the operation; and (3) that he had not sent cards to the physicians or senior surgeons asking them to attend the operation.

Pattison's twenty-one-page memorial of exculpation is the first of many florid statements he issued during his life. It is a long and bitter diatribe, and does nothing to enhance Miller's reputation. The tone is set by the opening sentence: 'Gentlemen: Egotism is always disagreeable to the hearer, and in this instance, believe me, most painful to the speaker.' He went on to say that his own character as a man and as a surgeon had been cruelly traduced. He then systematically dealt with the two surgical cases, and ended the memorial with a personal attack on Miller.

First he reviewed the facts about John Young. He stressed 'the unanimous opinion' that the femoral artery should be tied at the groin, and that this opinion was based on the view that the more serious operation of disarticulation was likely, in the end, to be necessary. He felt that he had in every way fulfilled the wishes of the consultants.

He then addressed himself to the accusation that he had performed the incision too high up on the thigh. He ridiculed the possibility of amputation four or five inches below the location of the femoral artery ligation, since gangrene of the stump would have inevitably followed: 'What! cut off the supply of blood from a part, and then call for an unusual and violent action from this weakened part?' He then discussed the likelihood of collateral circulation supporting the longer stump, and concluded that 'to medical men, at least to well-informed ones, the absurdity is so palpable as to be self-evident'. He finished his discussion of the Young case with the observation that 'had Mr. Rainy's opinion and mine been adopted, the patient would have had a more favourable chance of having his life saved'.

Next, he gave his views about Jean Gowdie's troubles. His initial opinion was that, in spite of every plan of treatment, she would die, either from 'overwhelming irritation' or from 'the tedious and debilitating process' that would be required for her recovery from such severe trauma. 'If we had one star of hope to look to, glimmering and feeble as its light might be, surely it was our duty as surgeons to assist nature to the utmost, to remove as much as possible the causes of irritation, and to convert as far as our art enabled us a complicated wound which could only be healed by a slow and weakening process into a simple one which if it healed at all would unite, as surgeons say, by the first intention, that is, speedily, with little irritation, and with hardly any exhaustion of the powers of the constitution.'

Pattison recalled that Miller, the first consultant to arrive, appeared to be quite friendly, and gave no indication of dissatisfaction with the operation on Young which had just been completed. But he seemed to have made up his mind in advance about Jean Gowdie, because he remarked to Pattison that 'he thought it was a pity that she had not got some artery opened, that he might make blood puddings of her'. After seeing her, Miller stated his positive opinion that nothing could be done for her, but Pattison averred that surgery might still save her life. Miller asked no one else's opinion and left the infirmary 'believing that nothing could be done'. Pattison found this astonishing, as Miller had already heard a contrary opinion stated and had seen Pattison's clerk, James

Armour, 'busily preparing the instruments' and asking questions about the mooted operation.

Pattison then asked emotive questions about letting one dogmatical and unphilosophical opinion regulate the practice of all the infirmary's surgeons, and about making no attempt to relieve 'a poor unfortunate young woman who had been the sole companion and the only support of a miserable, destitute, aged aunt'.

The operation went very well, according to Pattison, and it amazed him that very soon afterward derogatory whispers were heard regarding the two cases. It was to scotch these and to provide some background information that he then demanded a hearing before the managers.

At this point in his defence Pattison began a counterattack on Miller. First, he described an earlier case in the infirmary in which he, Pattison, had amputated a gangrenous leg. At the consultation prior to this operation, Miller, contrary to all the other surgeons, had stated that surgical intervention would be 'damned nonsense, damned butchery, or some such coarse expression', and that nothing could save the man. Miller continued by commenting that Pattison might butcher him if he pleased, and then cried out before the students, 'I shall wash my hands of his blood.' The amputation was successful and the man recovered; had he died, Pattison was to have been brought before the managers by Miller.

Next, Pattison described rumours that had recently been circulating about his 'cruelty, butchery and savage desire for cutting'. It was said that Jean Gowdie died not from the operation but from the manner in which it was performed; that, in fact, she had bled to death. Other rumours had it that he carried out the hip disarticulation only that he might claim to be the first man in Scotland to have performed such an operation. 'Mean, contemptible boast, worthy only of the grovelling soul who could engender it.' He then called on his colleagues, students, and patients to testify if they had ever had any reason to suspect him of inhumanity or brutality. 'If there is no voice of accusation, most assuredly I must call for my justification.'

Finally, Pattison stressed the damage that such rumours can inflict on the reputation of the infirmary itself ('this magnificent, this benevolent institution') by frightening off patients and discouraging benefactors. He added that such general damage was made worse by the uncouth behaviour of his adversary. 'Mr Miller, in a public party in the presence of ladies, positively advanced that the infirmary was now become a butcher's shop, and that amputation of two legs some few days before had been done, not that the removal of the legs was necessary, but that the surgeon wished it. What were females or strangers to think of this, coming from the mouth of a surgeon?' Pattison answered his own rhetorical question by predicting that no one would suffer his friends or pensioners to enter the infirmary nor would anyone contribute charitable funds for its future progress. He added that the two amputations alluded to by Miller had been correctly performed by 'that excellent surgeon Mr. Towers' after unanimous agreement had been reached at a properly constituted consultation.

'In truth, gentlemen, there are turbulent spirits, who change peace into anarchy, who convert order into turmoil and confusion; with such, even the most peaceable cannot live on terms unless they sacrifice their reason. I trust I shall be exonerated from the charge of egotism when this is considered, and I should fondly hope I have your warrant to say we find no fault with our surgeon.'

In spite of such eloquence, the committee's verdict on the two cases was 'that in operating upon John Young, Mr. Pattison did not comply with the directions of the consultation, whatever his own understanding may have been upon the subject; and that Mr. Pattison is not deserving of censure with respect to the case of Jean Gowdie, though the committee would have been better satisfied [had] Mr. Pattison, before performing so capital an operation as taking the arm by the shoulder joint, sent notice of his intention to the two senior surgeons of the House'.

The committee report was approved at a full meeting of the managers on 3 January 1817, at which Miller and one other followed up their advantage by insisting that the case of Jean Gowdie too gave cause for censure. Pattison was duly called before the meeting, when the Lord Provost of Glasgow read to him the conclusions of the committee's report and instructed him in future to 'attend implicitly to the regulations of the House regarding consultations and operations'. Finally, the managers on 6 January decreed by a majority vote that, in returning thanks as usual to the office bearers, Pattison's name be omitted. Miller was present at both of these meetings.

Pattison's friends and colleagues felt that the managers had been unduly harsh. Armour 'thought him to have been exceedingly ill used', and Rainy considered that 'upon the whole, Pattison had been treated, to say the least of it, with extreme severity'. Armour made some additional comments about the aftermath: 'Consultations since . . . have been exceedingly unpleasant. Mr. Pattison will not speak to Mr. Miller, nor will Dr. Graham.'

Miller's success in having Pattison censured by the board of managers was at best a Pyrrhic victory. His involvement in the Pattison affair in the dual capacity of director and senior surgeon raised a serious question of organization and administration. The board at its next annual meeting in January 1817 passed a resolution 'that from and after the first day of November next, it shall not be competent for any person to be at the same time a director of this institution and a medical attendant'. Miller resigned from the board on 6 October of that year; he died the following year, and the directors of the Glasgow Royal Infirmary paid tribute to his memory in the conventional phrases of institutional hypocrisy.

The quarrel with Miller with its humiliating termination was the first of many episodes in Pattison's life in which he offended the 'wrong' people, be they senior colleagues or prominent members of society. Of Pattison himself, at the time of the quarrel a twenty-five-year-old junior surgeon, it must be confessed that he was often over-confident, arrogant, abrasive, and insubordinate. In his encounters with senior and more conservative colleagues who made no secret of their hostility, clashes were inevitable. Compounding the effects of his naturally antagonistic disposition was a worrisome problem of communication between the medical attendants of the infirmary. In the case of Jean Gowdie, Miller claimed to have been unaware that a second consultation was to be arranged, while Graham disclaimed any knowledge that the first consultation had taken place. It is likely that Pattison, had he been so minded, could have averted these misunderstandings but deliberately refrained from doing so, since his chance of performing the shoulder disarticulation, which he felt was essential to save the life of the patient, would have been materially lessened.

It was the saving of life that was at the centre of this controversy. At that time the infirmary had been suffering a gangrene crisis. Overcrowding had augmented the spread of the disease, and Pattison must have been all too aware of the need for speedy radical surgery following trauma. Although one can feel respect and sympathy for Miller's wish that Jean Gowdie be 'allowed to die in peace', and recognize that major surgery subjected patients to excruciating agony, nevertheless Pattison probably acted correctly—according to contemporary standards—in his attempt to save the life of this unfortunate woman.

In the case of John Young, Pattison seems to have been determined to do the hip-joint operation, regardless of any opinion to the contrary. Such obstinacy occurred frequently throughout his life, suggesting a certain lack of imagination and insensitivity to suffering. But he may well have believed that injury to the femoral neck had been sustained, and that disarticulation offered the only hope for the life of the patient. If indeed he planned to complete the low thigh amputation, he erred in tying the femoral artery at the groin. The initial very high amputation virtually precluded a successful outcome; it is clear when one looks at the thigh one and a half inches below the groin crease, that it is impossible for a stump from a circular perpendicular incision to be covered with soft tissues and skin. It appears, then, that Pattison deliberately acted in defiance of the majority opinion. In short, there seems to be no reason to disagree with the verdict reached by the committee of enquiry.

Anderson's Institution, Glasgow

It is unlikely that Pattison was comfortable at the Glasgow Royal Infirmary after this unpleasantness. In any case, the regulations required that he go out of office after two years of service. Late in 1817 he completed his appointment as junior second-year surgeon and turned his attention to the future. While awaiting the results of his enquiries about a new position, he was delegated by the Faculty

Rival Lectures.

Mechanics' Institution.   *Anderson's Institution.*

Anderson's Institution. From the *Northern Looking Glass*. (Mitchell Library, Glasgow)
The Institution's avant-garde policies, which drew crowds of fashionable men and women to hear the lectures, contrasted sharply with the more down-to-earth approach of rival institutions like the Mechanics' Institute.

of Physicians and Surgeons of Glasgow to act as an 'examinator' and to prepare and transmit an address of condolence to the Prince Regent on the death of Queen Charlotte. Finally, he received word that he had been elected to the chair of anatomy and surgery at Anderson's Institution in Glasgow.

The Institution was founded in 1796, on the death of Dr John Anderson, as a rival to the University of Glasgow.[7] Anderson, formerly professor of oriental languages and natural philosophy at the university, was a quarrelsome bachelor with strong litigious tendencies. He had stipulated in his will that his estate be used to finance his proposed institution, at which the professors should 'not be permitted, as in some other colleges, to be drones, triflers, drunkards, or negligent of their duty in any manner of way'.[8] In life, he had encouraged the attendance of working-class people at his lectures, and liked to dispense with such formalities as the compulsory wearing of the scarlet undergraduate gown. In death, he achieved through the Institution two of his educational goals, both quite advanced for the time: the teaching of science to women and the popularization of science in the community by public lectures.

Anderson's estate in fact amounted to a debt of fifty-five pounds. The trustees he had nominated (one of whom was Pattison's father) could inaugurate initially only one lectureship, that in Natural Philosophy. The first incumbent, Dr Thomas Garnett, left an interesting contemporary description of the Institution in his book *Observations on a Tour through the Highlands and Part of the Western Isles of Scotland*. He was succeeded by Dr George Birkbeck, who supported Pattison in his later struggles at the University of London, and then by Dr Andrew Ure, who was to be Pattison's next antagonist.

A few years later was added the College of Medicine, for which funding was so short that the professors received nothing more than a title and tenancy of their lecture rooms. In applying for the Chair of Anatomy and Surgery, recently vacated by John Burns on his election to the Chair of Surgery at the University of Glasgow, Pattison no doubt considered the position a professional rather than a financial advancement. He continued to teach at the College Street Medical School throughout his association with Anderson's Institution.

His appointment was approved on 14 March 1818, but only on an annual basis; the managers and trustees who accepted his application must have been aware of Pattison's earlier troubles. He was formally called to a meeting of the managers to sign his name in the minute book, acknowledging this condition.

Following the confirmation of his appointment, Pattison left Scotland for France, where he studied in Paris for six months. At this time Paris had the undisputed leaders in medical research, who in turn attracted many outstanding postgraduate students from around the world. On his return in October 1818, Pattison resumed his old way of life in Glasgow, living at home with his mother and assuming his new teaching duties. On 20 November, his popular course of lectures on 'universal anatomy and physiology' at Anderson's Institution was advertised in the *Glasgow Herald*: 'In these lectures, Mr. P[attison] will demonstrate generally the structure of animal bodies and expose particularly the laws which regulate the animal economy. The introductory lecture will be open (*gratis*) to both sexes. . . . Fee: one guinea.'

Despite this promising start, signs of a fresh quarrel soon appeared when Dr Andrew Ure complained to the managers about the smell and nauseating remnants from the anatomy lectures. The problem was resolved by allowing Pattison the use of the main lecture room each evening, but requiring that he remove all his anatomical objects and preparations the same evening. Pattison was forbidden 'to introduce anything that might be considered by the managers as improper'.[9] But his troubles with Ure were only just beginning.

The Ure Divorce

Ure, some thirteen years Pattison's senior, was already well established at Anderson's Institution in the chair of natural philosophy. By the time Pattison was appointed, Ure was beginning to acquire an international reputation. He travelled widely, gave frequent and extensive lectures in the major centres, wrote papers and books including two dictionaries of science, and, whenever the opportunity arose, indulged in intemperate polemics against his fellow chemists. He is remembered now for his inhumane views on child labour, his antagonism to factory legislation, and his lyrical descriptions of factory life, which aroused the ire of the reformers.

In 1807 Ure married Catherine Monteath, who bore him two sons and a daughter. In the summer of 1818 she found herself pregnant again, but this time she claimed that the father was not her husband but Pattison. Ure never specifically denied paternity of the daughter born on 2 December; he simply accused his wife of adultery and sued for divorce.[10] Thus began another nasty scandal that was to dog Pattison for many years.

All proceedings connected with marriage (*vinculum matrimonii*) at this time were the responsibility of the Commissary Court of Edinburgh, which after the Reformation had succeeded to the jurisdiction in matrimonial causes previously exercised by the officials of the dioceses. Petitions for divorce involved only a 'pursuer' and a 'defender', the former merely filing a bill alleging that the latter was guilty of adulterous intercourse with a reputed paramour; no matter how innocent this last person might be, he was not a party in the cause and consequently might not appear in court. Clearly, collusion between husband and wife almost always resulted in the successful

Dr Andrew Ure. (Malloch Rare Book Room, New York Academy of Medicine)
As 'Dr Transit' he became the victim of the same satirical pen that had described Pattison as 'Beau Fribble': 'In the proud tone of conscious superiority, he intrudes on every company; he pesters the learned, and insults the ignorant, with impertinent recitals of his abilities, his merits and his labours. . . . Strange effect of science! that a man, as he excels in merit, should be more hated and despised in Society!'[11]

award of a decree of divorce. If no collusion had occurred, a second action usually followed: that is, the pursuer then sued the paramour for damages. The latter, having now become a party, could at last appear and have his say in court. It was at this point, then, that the pursuer and the paramour could meet in open legal battle. Only if a pursuer had any doubts about his case, or if some collusion had preceded the divorce hearing, would he not initiate this second action.

The story, which began in the winter of 1817, was revealed by the submissions of various servants and lodging keepers at a trial at which neither Pattison nor Mrs Ure was represented. Dr and Mrs Ure were a well-known and respected married couple, and Pattison frequently visited their house during that winter. He and Mrs Ure were found on several occasions to be standing together in a way that made the servants suspect that 'something was wrong between them'. This usually occurred when Ure was out lecturing. (Ure's lecture times were of course well known to Pattison.) Mrs Ure was described as 'flurried, not as usual' when observed unexpectedly in Pattison's presence by one of the servants: 'I observed Dr. Pattison and my mistress standing face to face, and close together; one of his arms was about her. When I opened the door, his arm fell down from her shoulder, her face dyed up red—she was in confusion.'

Many references were made to a back room in Ure's house, separated from the kitchen only by a thin partition of wood. Pattison and Mrs Ure were known to have repaired to this room on occasion. When the servants listened, they could hear 'stirring'. And on several occasions Pattison left his handkerchief in the house and returned for it later when Mrs Ure was alone. When Pattison then departed, she was seen to be 'confused and in tears'.

By August 1818 Mrs Ure's pregnancy was easily discernible. On the 12th, the Ures travelled by boat along the Forth and Clyde Canal to Falkirk, at that time a small country town midway between Glasgow and Edinburgh. After an hour or two, Ure left his wife in a boarding house and returned to Glasgow.

Two days later, on 14 August 1818, she wrote a letter to Pattison, then working in Paris. She also sent a copy of the letter to Ure! Her motives for doing this became the subject of much discussion. Pattison claimed that he never even received the letter, which he believed was simply part of the Ures' collusion. Ure, on the other hand, contended that his wife, 'being in an unhappy state, made a full declaration of her guilt'. The letter, a bitter, melodramatic denunciation, must have had its effect on the judges:

> With a mind overwhelmed with grief, and a breaking heart, I again sit down to address you; it is to me a task of the most painful kind, but my forlorn and destitute situation calls loudly for you to come and give me relief. Oh! Granville, will nothing awaken your feelings or compassion towards me? Must I die here in misery and want, without one consoling word from you, the author of all my misfortunes? . . . And allow me to ask you, what is to become of the innocent offspring that may be looked for in a short time? I am now five months and a half gone with child to you. . . . The child you must take, as you well know you are the undoubted father of it. . . . You have brought me into this awful state, and it is to you alone that I can look for support.

Mrs Ure lived in her Falkirk lodgings for two months, answering to the name of Mrs Campbell. She was treated miserably, as she described in a letter to a friend, Mary Park: 'You cannot imagine the insolence I received from these two wretches [the lodging house keepers]. . . . When I got a little bit of meat, they both instantly examined it and would have taken as much of it as they thought decency would admit, and everything in the same way.'

She moved to Edinburgh in mid October, about six weeks before her confinement. Her shame made her resort to subterfuge. 'I came in a post chaise to the end of Princes Street; I shifted my trunk into a coach which I got on the stand, paid the driver and returned him. This is all for safety.' Her new landlady and lodgings were much more satisfactory than those she had left in Falkirk. The landlady, after some hesitation, even gave permission for the delivery to proceed in her house. Mrs Ure had now become 'Mrs Thompson, a lady from the country, in bad health, come to be near medical assistance'. Despite these improved circumstances, her misery and loneliness were unmistakable. 'It is now more than ten long weeks since I left my beloved family. How many days and nights have I spent since that time of the most poignant sorrow. . . . What would I give to see them: they are never from my mind, day nor night, and my dear boy to be ill, how unfortunate when I was from home.'

Mrs Ure gave birth to a daughter on 2 December 1818, and on the very same day she was served with a summons for divorce. Ure was ordered to name the adulterer referred to in the summons. This he did in a condescendence (statement) dated 8 January 1819: 'The pursuer verily believes the adulterer to be Granville Sharp Pattison, Surgeon in Glasgow.' On the same day, Ure, 'kneeling with his right hand on the Holy Evangel', swore an oath of calumny, in which specific denial was made of any concert or collusion between him and the defender.

The formal proceedings began on 30 January 1819 and continued intermittently for several weeks. The commissaries produced an interlocutory (provisional) judgment on 5 February 1819, finding Mrs Ure guilty of adultery with Pattison. At this time he claimed to have been unaware even that Ure had initiated any divorce proceedings. On learning of the interlocutory judgment and of his reputed role as the paramour, he immediately instructed his solicitor to see Mrs Ure. She was persuaded to sign a mandate for appeal denying the charge of adultery and claiming that Ure had continued to

The first page of Andrew Ure's condescendence in which he named Pattison as his wife's paramour. Scottish Record Office CC 8/6/117. (Published with the approval of the Keeper of the Records of Scotland) cohabit with her even after her banishment to Falkirk. Proof of cohabitation would have amounted to condonation and would have barred Ure's claim for divorce.

Some three weeks later she reversed her stand by disclaiming the terms of the mandate, disavowing the accusation of cohabitation levelled against Ure, and offering no opposition to the divorce. All obstacles were removed for the decree (final judgment) and sentence. On 26 March 1819, Mrs Ure was found guilty of adultery with Pattison and was accordingly divorced from her husband.

At this point Pattison might have sued Ure for defamation. That he failed to do so casts some doubt on his innocence. Instead, he attempted to clear his name, first in Scotland by preparing 'A Statement of Facts' illustrated with facsimiles of Ure's letters (this was legally suppressed by the Chancery Court on the morning of its announced publication on the ground that Pattison did not have the authority to publish Ure's personal letters) and later in the United States by circulating two very lengthy and bitter pamphlets, *A Refutation* and *A Final Reply*.

First, he presented Ure as a 'degraded and infamous character'. He cited Ure's unsavoury reputation in Glasgow, where he was 'shunned and despised', and followed this with a charge that Ure had only just escaped transportation to Botany Bay for fraudulently stealing and destroying his father's will. (In fact, Ure had so enraged his father by not entering the family business that he was cut out of the will; whereupon he showed his displeasure by wresting the will from a lawyer and flinging it into a fire. Ure could have been found guilty of theft or malicious damage.) Pattison then described Ure's devious behaviour at the time of the divorce trial: he had given no indication to Pattison that the action had even been initiated and indeed had behaved toward him with obsequious friendliness. Finally, he accused Ure of having offered his wife a sizeable annuity if she would not contest the divorce action.

Next, he mounted an attack against Mrs Ure. He pictured her as a poor, weak, vacillating creature; in total, she had made four statements under oath, two averring Pattison's guilt and two his innocence. He stressed her duplicity and lies and suggested that this obvious perjury stemmed from real or imagined bribes or promises made by agents acting for her husband or himself.

The Ures were then jointly accused of gross moral turpitude in connection with a lascivious letter from Ure to his wife, dated 12 October 1818. Described as 'execrably obscene' and 'detestably indelicate',[12] it was written at about the time Mrs Ure moved from Falkirk to Edinburgh and more than two months after Ure had banished her from his company. The letter made it clear that Ure had not only known of his wife's affair, but was prepared to condone it: 'You may write to P. that I shall meet him here without hostility if he promises to put a more careful restraint on his machinery for the future; or at least if he and you don't conspire to deceive me, but behave openly and honourably in your love affairs.' His only regret, 'that (as Mrs P[ark] says you are very big) a person on visiting you in Edinburgh, as I may do, will not be able to come near your centre of attraction', was hardly the reaction of an outraged and unforgiving husband. Pattison claimed this as further evidence of the Ures' collusion, particularly as the letter indicated that they were writing regularly to each other at this time. Ure denied authorship and accused Pattison of having forged it; his accusation was supported by Mrs Ure in a signed affidavit.

Pattison then went over the records of the trial, point by point. He first stressed that all the evidence was circumstantial and that the examinations were *ex parte* and unchallenged. He then pointed out that the evidence was derived from two sources: disinterested witnesses, such as servants, and Mrs Ure herself. The testimony of the first, if divested of presumptive colouring, provided no evidence of sexual intimacy but merely confirmed that he was in the habit of visiting the Ures. In proven cases of adultery, servants had brought forward unequivocally damning proof, but in this case he considered the evidence to be 'so inconsequential as to be exculpatory'. As for Mrs Ure's testimony, he had already proved it to be worthless.

Considering the letter of 14 August 1818 written by Mrs Ure ('with a mind overwhelmed with grief'), Pattison stressed the monstrous absurdity of her providing Ure with written evidence of adultery if she had indeed been guilty; but it was this evidence, and it alone, which appeared to incriminate him. He regarded the letter as a mere link in the chain of collusion and conspiracy and even suggested that Ure had written it for his wife to copy.

The charge of paternity was, to Pattison, clearly untenable. How could Mrs Ure have known that he was the father of the child? It had been admitted that at the time of its conception the Ures had been living together in all the intimacies of the married state; had she at the same time been carrying on an intrigue with another man, still she could not possibly have known who was the real father.

To conclude his defence, Pattison tried to picture a hypothetical course of events which might explain Mrs Ure's totally unpredictable and extraordinary behaviour. He pointed out that in the summons she had been accused of having committed acts of adultery in June and July 1818 at the Ures' holiday residence. As Pattison had been in Paris during the whole of that summer, these acts must have been committed by someone else. He offered the theory that Ure, having discovered his wife in an adulterous relationship with 'some servant or some inferior person', compelled her by threat of exposure to conspire with him in identifying Pattison as her paramour. This would have lessened her disgrace and might have enabled Ure to obtain damages.

Many months after the divorce had been granted, Mrs Ure finally confessed that Ure's lewd letter of 12 October 1818 was indeed the work of her quondam spouse and that she had lied when she stated in her affidavit that the letter was a forgery. The reason she gave for disclosing her perjury was that a bribe offered by Ure for signing the affidavit had not been paid. This gave Pattison the opportunity in October 1819 to institute an action against Ure for £2000 on the charge that he had been falsely accused of forgery. The suit was resisted and then withdrawn because of Pattison's absence from Scotland.

It was not proven that Pattison and Mrs Ure were lovers, nor that he was the father of the child. She seems to have been a weak, compliant, opportunistic woman, highly susceptible to bribery and threats. Blown this way and that, she perjured herself all the while. What is certain is that Pattison made no provision for her or the child in his will. Ure himself appears to have been a hard-working but unlikeable man, who saw the opportunity of release from an unsatisfactory marriage and of damaging the reputation of an arrogant colleague. His failure to sue Pattison for damages following the successful divorce proceedings is at least consistent with collusion with his wife. His character and reputation were cast into doubt by the perceived depravity of his letter of 12 October and by publicized instances of lying and bribery. Ure resigned from Anderson's Institution in 1830 and became a successful consulting scientist in London.

As for Pattison, there can be no doubt that he was indiscreet in his frequent visits to another man's wife, particularly in an age when such conduct was judged much more severely than it is now. The verdict of his colleagues was harsh. 'The impression of his guilt is so strong that most of his friends do not associate with him.'[13] John Robertson, Pattison's friend and colleague at the College Street Medical School, 'was quite convinced not only that Pattison was guilty but that he had been perfectly informed of all the legal proceedings while in progress'.[14] The original papers of the trial and the subsequent refutations show no convincing proof of his guilt; at best the evidence is inconclusive. As Pattison was at no time a party in the dispute, a judgment against him under the law of the Commissary Court was never at issue; the judges considered him guilty by implication. A more reasonable lay verdict would perhaps have been the by now familiar 'not proven'.

During these difficult days in the winter of 1818/19, Pattison was still lecturing as usual. In addition to his position at Anderson's Institution, he spent most of his time with a large class of students at the College Street Medical School. In late March 1819 he told the class his side in the Ure divorce proceedings and read aloud Ure's letter of 12 October. A few days later, on 6 April 1819, his College Street students presented him with a ring and a letter:

Sir,
The gentlemen attending your Course of Lectures on Anatomy and Surgery beg leave at this time, when the present session is drawing to a close, to present you with a diamond ring, as a small testimony of their high opinion of your talents; of their gratitude for your exertions; of their sense of your obliging, liberal and gentlemanly conduct on all occasions;

and of their heartfelt wishes for your future happiness and prosperity.[15]

(One of the co-signers of the letter was John Conolly, who was to provide support for Pattison again in 1831 during the final phase of his battles at the University of London.)

The divorce became a topic of urgency for the managers of Anderson's Institution. On 15 April 1819 they were given the Extract Decree of Divorce for study and on 6 May were unanimously of the opinion that it demanded their immediate attention. Pattison was summoned to appear before them on 13 May. Instead of obeying the summons, he saved the managers considerable trouble by resigning his chair of anatomy and surgery on 8 May:

> I had hoped when the managers did me the honour to appoint me, that I should long have continued to have fulfilled the duties of that most respectable situation, but as I cannot bear that my name should be associated with Dr. Ure's in any institution, I have long ago resolved to make this sacrifice to my feelings. . . . It may perhaps be supposed that my resignation has arisen from my unwillingness to enter into any investigation of this business. If my conduct throughout the affair be reviewed, it will be observed that I have never courted concealment.[16]

He pointed out that the interlocutory judgment was obtained by 'the most shameful collusion between Dr. and Mrs. Ure' and that the decree was procured only by 'bribery and perjury'. He then requested a public meeting of the managers and trustees to hear his side of the controversy.

In considering the letter, the managers observed that Pattison had not denied the charge of adultery with Mrs Ure. They therefore took no action regarding his request for a public hearing and recorded that his resignation had rendered it unnecessary to pronounce a sentence of expulsion. 'Yet they deemed it necessary to declare their abhorrence of his conduct and that he is unfit and ineligible ever to hold any situation in the Institution.'[17]

Pattison's career at Anderson's Institution had ended abruptly and ignominiously after only fourteen months. It is worthy of mention, as an ironic footnote, that the man appointed to succeed him, Dr William Mackenzie, was soon embroiled in controversy with Ure over lecture rooms, for precisely the same reasons that had caused Pattison's original rift with Ure.

Pattison's last days in Glasgow were miserable. Thomas Lyle described them in his journal of 4 March 1819:

> The principal topic of discussion here for some time past has turned upon the bankrupt character of poor Mr. Pattison. Woe to the day that stamped him down among the sins of folly and libertinism. . . . Mr. Pattison has been gradually sinking into the nadir of popularity, in spite of his every exertion to the contrary; his medical practice is vanishing from him, genteel society avoids his presence, while his medical friends are deserting him to a man. He has lost his lectureship, and now wanders about like Cain, with the curse upon his forehead.
>
> Lately, when the Faculty had a meeting and a full assembly of its members were present, when they found they had some business to arrange in Edinburgh, Mr. Pattison, being present, volunteered his services as he required to be there himself soon. 'No,' retorted Dr. [John] Nimmo [past president of the Faculty], 'your private character at present lies under a stigma, and by employing you, our own good name would be abused.'

Departure from Scotland

Pattison was only twenty-eight years old yet had already suffered more than his share of troubles. A crisis had occurred during the tenure of each of his three positions in Glasgow: the grave-robbing trial at the College Street Medical School; the Miller quarrel at the Glasgow Royal Infirmary; and the damaging charge of adultery in the Ure divorce trial at Anderson's Institution. These episodes all contributed to his increasing obstinacy, aggressiveness and embitterment. Only during the last few years of his life did his character show signs of softening.

On 22 May 1819 Pattison quietly left Glasgow, bound for London and Philadelphia and a fresh start in life. During the difficult months of the Ure divorce proceedings he had been planning his departure, but he had not divulged his plans to anyone. Throughout the summer concern and uncertainty about his intentions had been expressed in the letters of his colleagues in Glasgow: 'Anyone who had any idea of lecturing in Glasgow would do well to ascertain certainly whether he meant to return or not. I do not think Pattison will ever succeed as a surgeon in Glasgow after what has happened. I am not so sure it would materially injure him as a lecturer. The students seem to like him.'[18]

He reached London on 24 May, where he remained for five days only. He made calls on many of the most eminent London anatomists and surgeons, including Mr (later Sir) Astley Cooper, from whom he received letters of introduction to colleagues in the United States. Next, he set about increasing his qualifications and managed to be elected to two august organizations during those five days.

First, he approached the Medical and Chirurgical Society of London. This organization was founded in 1805 by the chief London physicians and surgeons of the time, including several of the men whom Pattison visited. (The society received a royal charter in 1834 and later became a constituent part of the Royal Society of Medicine in 1907.) The process of election to its membership has always been that of a club: a candidate is proposed and seconded, his name goes to ballot at a general meeting, and if there be no opposition his election follows. It is likely that Pattison had done some preliminary groundwork from Glasgow, and that his reputation and character were still unsullied amongst the London medical fraternity. He was duly elected to the society and retained his membership for the rest of his life.

His second approach was to the Royal College of Surgeons of England. Election to membership was by examination but this was probably not stiff as there are scarcely any recorded rejections. The examination, which followed the usual formalities of registration and submission of supporting documents, took place on a Friday evening at six o'clock. The president took the chair, and nine members of the court occupied seats around a horseshoe table, with the candidate sitting at the bottom. Questions were asked by all members and covered a wide range of disciplines, including anatomy, general medicine, physiology, and surgery. Pattison passed and was elected a member. (He later claimed, incorrectly, to have been elected to a Fellowship. The first Fellows were created only in 1843.)

He left London on 28 May 1819, secure in the knowledge that he had acquired two valuable new qualifications and letters of introduction from many of London's medical elite. He arrived in Liverpool on 30 May, and embarked the following day in the packet ship *Courier*, bound for New York.[19]

IV
Teaching Successes and Quarrels in the United States 1819–1826

IT TOOK PATTISON about five weeks to complete the Atlantic crossing, landing in New York on 6 July 1819. His brother John and sister Margaret were living in America at the time—the family's muslin works in Glasgow had failed some years earlier and John had temporarily forsaken Scotland to become a Philadelphia merchant. The day after his arrival, Pattison boarded the stage for Philadelphia, where he was welcomed by John and Margaret that evening. After spending about an hour at John's house, he was taken at once to meet some of his future medical colleagues. The meeting was very cordial, but Pattison learned that problems were developing regarding his future employment. Soon he was to become the central figure of one of the most colourful feuds within the nineteenth-century American medical profession.

The idea of moving to the United States had been first mooted in a letter from John which Pattison had received on 24 December 1818. (At this time Pattison was the newly appointed professor of anatomy and surgery at Anderson's Institution, and the legal proceedings in the Ure divorce case were already under way.) John's letter stated that the incumbent professor of anatomy at the highly regarded University of Pennsylvania had died suddenly, and that Pattison could be elected to fill the vacancy. Dr Philip Syng Physick, the professor of surgery, had volunteered in the meantime to carry on the class of anatomy in addition to his regular duties. John mentioned a probable salary of '$8,000, nearly £2,000, besides your chance of practice'. He urged his brother to procure strong letters of recommendation from as many eminent medical men as possible. These were duly forwarded without delay to the trustees of the university along with a formal application dated 1 January 1819.

On 17 May 1819 Pattison had received a letter from Dr William Potts Dewees, adjunct professor of obstetrics at the university.[1] One sentence in this letter became the focus of much subsequent argument: 'I have no hesitation to declare, that no question remains in my mind, that were you on the spot, your election would be certain.'[2] In spite of these encouraging words, Pattison was less than confident about his election to the coveted chair, but, being out of work and out of favour in Glasgow, he immediately made plans to travel to Philadelphia at the earliest opportunity, with the intention of giving private lectures there if he did not get the professorship. His fears were well-founded: on 6 July 1819 his application was rejected by the trustees of the university; and one day before his arrival Physick was persuaded to accept the chair of anatomy permanently. This was undoubtedly a great blow to Pattison. Apart from his brother and sister, he was alone in a new and strange country, with no friends to console him. However, he did find some consolation, and indeed encouragement, from Nathaniel Chapman, who was to play a major role in Pattison's life in Philadelphia.

The Chapman Quarrel

Born in 1780 into a comfortable, upper-class home, Nathaniel Chapman began the study of medicine at the age of fifteen as an apprentice and later as a private pupil of Dr Benjamin Rush, one of the original signatories of the Declaration of Independence. (Chapman subsequently quarrelled with Rush in an unseemly manner.) Chapman had been elected to the chair of materia medica at the University of Pennsylvania in 1813 and to the prestigious chair of the theory and practice of medicine in 1816. One year later he founded the Medical Institute of Philadelphia, which offered

Nathaniel Chapman, professor of the theory and practice of medicine at the University of Pennsylvania. (Malloch Rare Book Room, New York Academy of Medicine)

supplementary lectures to students of the university; the programme proved to be very popular. The students held him in high
regard as a teacher—after they had become accustomed to his
defective enunciation resulting from a cleft palate—but they were
aware of his unwillingness to accept new ideas and his advocacy of
out-of-date concepts. He was well-connected socially, ambitious,
proud, and powerful in local medicopolitical manoeuvrings. To
cross swords with him boded ill, as Pattison was soon to discover.

Chapman initially tried to ease Pattison's disappointment over his
failure to be elected to the chair of anatomy by suggesting that,
instead, he would be likely to be elected to the chair of surgery, just
vacated by Physick. However, soon afterwards, Dr William Gibson,
professor of surgery at the University of Maryland, was appointed
instead. (Chapman, it transpired, had supported both Physick's and
Gibson's elections to Pattison's detriment.) Chapman then suggested
that Gibson and Pattison should share the surgical chair. But when
Gibson, on arrival, argued that student fees would be inadequate to
support two professors, the plan was abandoned.

Concerned about his future, Pattison asked Chapman, as a friend,
for a frank appraisal of his prospects in the United States. Chapman
replied that he himself was anxious to resign from the Medical
Institute of Philadelphia. He added that the new incumbent would
be made a professor of the university, and that Pattison's election
was certain. Yet, shortly thereafter, Chapman hinted that the trustees
might find it impossible to confer this professorship on Pattison
'because he was a foreigner'. Sensing that he had again been misled
by Chapman, Pattison was furious and expressed his views with
considerable feeling, whereupon Chapman 'had the indelicacy to
propose that I should associate myself with Dr. Physick's dissector'.[3]
Pattison was astonished and insulted by this latest suggestion, since
he considered his standing fully equal to that of Chapman. He
decided, nevertheless, to stay in the United States for at least a year;
if no university appointment were forthcoming, he would give
private lectures on anatomy and surgery as originally planned. This
decision, doubtless based on a combination of pique, anger, and
obstinacy, was more or less forced upon Pattison: he had previously
sent instructions to Glasgow, apparently on Chapman's advice, that

his valuable museum of anatomical and pathological specimens should be shipped to him in Philadelphia; this was now in transit.

On 4 October Pattison duly requested that the trustees grant him permission to house his museum within the walls of the university and to deliver a popular course of lectures on anatomy and physiology for two evenings a week in one of the unoccupied classrooms. In return, he offered the medical professors the free use of his museum for illustrating their lectures, so long as the preparations remained solely under his control. The board referred the requests to a committee of three, which recommended that they be approved. But the board overruled the committee and instructed the secretary to inform Pattison that the standing rules of the institution did not permit them to grant his requests. One of the board members on this occasion was General Thomas Cadwalader, Chapman's brother-in-law.

The museum arrived safely and Pattison started a course of lectures. The location of both museum and teaching rooms remains uncertain. Chapman claimed that he and Physick prevailed on the trustees to allow the museum and the lectures to be accommodated in the university; however, the minutes of the trustees contain no entries to support this contention. Chapman and Physick attended Pattison's introductory lecture, and he reciprocated by hearing all the introductory lectures in the medical faculty. Even at this stage, the impending petty squabbles might have been avoided had not new infractions of etiquette, real or imagined, initiated further episodes of irritation and resentment.

Pattison had been careful to select a time for his private lectures that did not conflict with the scheduled courses of any of the medical professors. Yet no sooner had he started his lectures than Chapman arranged for his private class at the Medical Institute to meet at exactly the same time. Three weeks later, Pattison began a second course of lectures, on the subject of surgical anatomy. Chapman made further changes in the time of his private class to coincide with these too, thus effectively preventing class members from attending any of Pattison's lectures.

From this point on, all friendly intercourse ceased.[4] The first public show of hostility occurred at a meeting of the Medical Society of Pennsylvania, of which Pattison had been elected an honorary member. Chapman had ruthlessly crushed a young physician who had 'dared' to differ with him. Pattison rose to support the victim, whereupon Chapman attempted to overwhelm Pattison with invective. He ended by observing 'that Americans were not to be instructed by a foreigner, an ignorant humbug'.[5]

It was at about this time, in the autumn of 1819, that Pattison received an offer of the professorship of anatomy at the Transylvania University of Lexington, Kentucky. He claimed that the offer, unsolicited, came only after he had completed all arrangements to lecture in Philadelphia, and that consequently he declined it. Chapman contended that Pattison had actively sought the appointment, in order that it might be published as an early recognition of his consequence in the United States, and that the trustees at Lexington were justly indignant at the disrespectful manner in which they had been treated. He stated that Pattison made his application through Dr Charles Caldwell, a professor at the Transylvania University at that time. But Caldwell did not refer to any such incident in his outspoken autobiography. A man who never forgave or forgot a slight, he made no mention of Pattison, which suggests that Chapman's contention of Pattison's duplicity was unfounded.

The feud with Chapman simmered throughout the winter of 1819/20, with further petulant outbursts. Then in August 1820 Pattison was unexpectedly offered the chair of surgery at the University of Maryland in Baltimore; this was the position vacated by Gibson the previous year. Pattison moved almost immediately to Baltimore, where he hoped to be free from all the rancour of the winter just past.

Even after Pattison's departure from Philadelphia, Chapman continued to vilify him, calling him 'a Scotch blackguard, a refugee driven from his country'. The quarrel climaxed when Chapman publicly accused Pattison of writing him an anonymous letter. Characteristically, Pattison wrote to Chapman on 12 October 1820, demanding to know if he had made the accusation and 'seeking that redress, which every gentleman is entitled to demand and no-one, if he has injured another, can refuse'. A follow-up letter was necessary

before Chapman replied—to an intermediary—on 19 October: 'I
have determined to hold no communication with [Mr Pattison], by
correspondence or otherwise. If any further proceedings on the part
of Mr P. should render it necessary, I shall take an opportunity,
through the medium of the press, of assigning the reasons which
have led me to this decision.'

Pattison immediately travelled from Baltimore to Philadelphia
and posted broadsides in two public places:

TO THE PUBLIC

Whereas Nathaniel Chapman, M.D., Professor of the Theory
and Practice of Medicine in the University of Pennsylvania has
propagated scandalous and unfounded reports against my
character; and Whereas when properly applied to, he has
refused to give any explanation of his conduct, or the satisfaction which every gentleman has the right to demand, and
which no one having any claim to that character can refuse, I
am therefore compelled to the only step left me, and post Dr.
Nathaniel Chapman as a Liar, a Coward, and a Scoundrel.

Granville Sharp Pattison

Philadelphia, Oct. 23d, 1820

Within an hour or two, Pattison was arrested for the offence of
posting at the instance of one of Chapman's brothers-in-law. Information about subsequent events is sparse. He was greeted with
sneers by Chapman's friends on his arrival at the mayor's office. He
was later examined before a grand jury, who rejected the indictment,
apparently at Chapman's request.[6] He was therefore released but
was bound over to keep the peace for a period of two weeks, on his
recognizance of $14,000.

Chapman had not accepted the implied challenge of a duel;
Pattison later upbraided him for his cowardice and branded him as a
braggart for his references to 'ten paces' and 'the blowing out of
brains': 'I did not suspect when I wrote to Dr. Chapman that my
letter was to verify the fable of the ass with the lion skin; I never
suspected that he was a lion, but I have long considered him a
serpent.'[7]

As reasons for refusing Pattison's challenge Chapman had cited the disparity in their ages, the inequality of their social standing, the obligations of decorum imposed by his professional position, and the claims of his numerous family. '[Mr Pattison] is an adventurer with a tainted reputation which he hoped to repair. What has he to lose? To ruin the happiness of a family, we have already seen, is one of his sports.'[8] With his customary eloquence Pattison ridiculed Chapman's reasons. Regarding the age difference: 'I never understood that the laws of honour allowed a man to insult another with impunity, simply because he is twelve years older.' Always keenly aware of his own position, he gave no ground on the question of social standing: 'If we come to measure the line of our ancestry or the eminence which we hold in our professions, I surely am not second to Dr. Chapman.' As for professional decorum: 'I believe that neither the parents nor friends of the several hundred young men committed to our care would have wished us, because we were professors, to bear insult unnoticed.' Finally, Chapman's concern over his family responsibilities elicited little sympathy from Pattison: 'I have not a numerous family, I admit; but did Dr. Chapman, before he had a family, when he challenged Dr. Dewees,[9] think that this excused him?'

The duel of honour, which recurs throughout this book (as indeed it does in the history of those times), was a subject upon which Pattison held strong views. (The decline in popularity of 'pistols for two and coffee for one' after about 1850 may realistically be attributed to the greatly improved accuracy of the hardware. It became expedient for a man to swallow his pride when muzzle loaders and hand-packed powder gave way to deadly hand-guns.) Although Pattison disclaimed being a professional duellist, his attitude to duelling was firm: 'There are two codes of laws under which all men of principle and honour must be ranked. I allude to those of religion and honour.' If a man lived a truly devout and humble life, and if he bore with meekness unmerited insult, 'we know that he is not a coward, but a Christian'.[10] He admitted that his general conduct did not entitle him to shelter under the laws of religion, consequently he felt impelled to be guided by the rules of honour.

The Pamphlet War

It was about two weeks after Pattison's arrest and release that the pamphlet war with Chapman began. Four pamphlets, 200 pages in all, were written in 1820 and 1821. Widely circulated over the continent, they contained a barrage of charges and countercharges. The narrative was complicated by references to many minor characters who surfaced briefly in a variety of little episodes. Much of the writing was bombastic, laborious, and inaccurate, and many of the points were contradictory and unverifiable, due to a scarcity of collateral evidence supporting one side or the other. Stylistically, Pattison's pamphlets were superior to Chapman's, but this does not imply that they were more truthful. This exchange, which illustrates a common means of venting spleen at that time, had little impact on anyone but the two protagonists.

The first salvo, fired by Chapman on 5 November 1820 in his *Correspondence between Mr. Granville Sharpe Pattison and Dr. N. Chapman*, retold the tale of Pattison's initial quarrel with Chapman. In it Pattison was depicted as an adventurer and fugitive from Scotland and his scandalous past. Far from acknowledging that he had encouraged Pattison's application, Chapman portrayed him as an unwelcome upstart who, to Chapman's surprise, had arranged for unsolicited testimonials to be sent to Philadelphia from Britain. When, on arrival, Pattison had been informed that the anatomy chair had been filled, instead of following Chapman's advice to return to Britain or to apply for the chair of surgery at the University of Maryland, he had had the presumption to inaugurate a course of private lectures in rivalry to those at the university. To augment his attack, Chapman, convinced of Pattison's guilt, dwelt at great length on the Ure divorce and the depravity of the alleged paramour. Thus was the groundwork laid for full-scale war.

Pattison was not slow to respond. Just three weeks after the publication of Chapman's *Correspondence*, there appeared *A Refutation of Certain Calumnies published in a pamphlet entitled 'Correspondence between Mr. Granville Sharpe Pattison and Dr. Nathaniel Chapman'*. There is a similarity of style between the opening lines of this and those of his memorial of exculpation at the Glasgow Royal Infirmary:

> To be forced to appear before the public, even in defence of professional reputation, is exceedingly painful. To be obliged to come forward and defend moral character is a thousand-fold more so. I am a stranger in this country, and it has unfortunately happened that, ever since my settlement, I have been engaged in controversy and disputation.

Pattison described at length his reasons for moving from Glasgow to Philadelphia and the treatment he had received upon arrival. He expounded on Chapman's main accusation: that he, as the guilty party in the Ure divorce, was hounded from Glasgow in disgrace to take refuge in the United States. He refuted Chapman's assertions by the quotation and interpretation of a variety of letters, signed depositions, certificates, and legal opinions.

Stung by Pattison's spirited counterattack, Chapman, nine months later, provided two further pamphlets. The first, dated August 1821, was a second and expanded edition of his *Correspondence*. This was followed on 1 September by the publication of the full records of the Ure divorce: *Case of Divorce of Andrew Ure, M.D. v. Catharine Ure*. In this, Chapman commented freely on the court proceedings and called on Pattison's friends and colleagues 'to mark the perfidy practiced on them'. Chapman then had this to say about the quarrel:

> That I have not been actuated by any vindictive motives, in this case, my heart assures me. The controversy was of his own seeking, and most reluctantly did I approach this huge mass of moral putrefaction—being fully aware, that it was not to be stirred without contamination. Could I have discerned, even with the strong provocations I had received, a spirit of contrition, or of ordinary decency and moderation, in the habits and deportment of that individual, I think I should have remained silent. But in place of this, finding that his insolence increased in proportion to my forbearance, and that an immunity from punishment seemed to promise only a repetition of crime, I felt it incumbent on me to act as I have done.

Pattison produced the last pamphlet of the series some time after the appearance of the court records: *A Final Reply to the Numerous Slanders circulated by Nathaniel Chapman, M.D.* Freely spiced with invective and insults against Chapman, the pamphlet presented a convincing resumé of the various squabbles and seemed to invalidate many of Chapman's accusations. Pattison ended with a reference to Chapman and his friends: 'I have now forever done with these men. . . . Should they ever hatch up any other calumny, I shall treat it with the silent contempt it merits.'

One further episode involving Chapman occurred on 7 May 1821, at the height of the pamphlet war, when he encountered Pattison unexpectedly on a Philadelphia street. Chapman, who was walking with his wife and a brother-in-law, crossed the street and struck Pattison with his stick. Pattison fought back, and the affray ended only when Mrs Chapman screamed and fainted. Pattison informed the public of Philadelphia in a Baltimore newspaper that Chapman could at all times feel fully protected, 'if in a lady's presence'. Chapman, for his part, fully acknowledged that he had dealt Pattison several blows; he publicly expressed regret, and explained his behaviour by a 'momentary heat' and 'sudden, irresistible excitement' on seeing unexpectedly his hated adversary.

Chapman was not personally involved in the final round in the battle, when, two years later, Pattison fought General Thomas Cadwalader, Chapman's brother-in-law, in a pistol duel. The two men were never reconciled, even when Pattison returned to Philadelphia ten years later. It has been observed that had the duel been between Pattison and Chapman, hostilities might have diminished, but when Chapman allowed the general to suffer in his cause, 'all approximation of the original contending parties became impracticable'.[11]

The Chapman–Pattison quarrel was basically a personality clash between a well-educated, socially prominent, older member of the Philadelphia medical establishment and an irrepressible, opinionated young foreigner. Pattison's legal and moral difficulties in Scotland made him an easy target. Chapman looked upon him as a pariah, a ruined man trying to escape the devastation of Presbyterian social outrage.

Why did Chapman and his medical colleagues gradually turn amity into enmity? It was probably because biased reports from Glasgow were filtering in about the Ure divorce and the earlier grave-robbing scandal. Both Pattison and Chapman indulged in intemperate polemics, freely laced with half-truths and outright falsehoods; both were very much alike: intelligent, self-centred, proud, and dignified. '[Chapman] was "the prince of good fellows" as long as he was not crossed, but when disputed, or if he felt his professional position in jeopardy, he could demonstrate a marked viciousness, a supreme accomplishment in the art of backbiting, and a truly amazing degree of flexibility in beliefs and basic tenets.'[12]

The Gibson Controversy

The pamphlet war with Chapman took place after Pattison's move to Baltimore in August 1820, but even before this he had taken up arms on another front, this time against William Gibson, the newly appointed professor of surgery at the University of Pennsylvania. The quarrel arose over a fascia (a sheet or band of fibrous tissue) which Pattison had demonstrated but had mistakenly claimed as a new discovery; it subsequently proved to be Colles's fascia (the superficial fascia of the perineum) which was already known.

It was said that Gibson was not an amiable man. Vain in his personal appearance, he was even more so in his profession as surgeon and teacher. Yet his style of lecturing was easy, agreeable, and instructive, and he commanded a loyal following among the medical students.

On arrival in Philadelphia in July 1819 Pattison had announced the discovery of the new fascia which, in the ensuing months, he demonstrated to many of the Philadelphia anatomists and surgeons, who allegedly were excited and impressed. Eventually the findings were described in January 1820 in a paper entitled 'Experimental Observations on the Operation of Lithotomy, with the description of a fascia of the prostate gland which appears to explain anatomically the cause of urinal infiltrations and consequent death'.

In this article, Pattison described some of his surgical experiences in removing urinary calculi from the bladder. At that time, the operation of lithotomy was performed by a large lateral incision through the perineum and prostate gland. He concluded that death usually resulted from urinary infiltration into the space between the bladder and the rectum, with subsequent suppurative infection. He gradually realized that a small incision, performed above the base of the prostate, gave a greatly improved prognosis. He concluded that there might be a fascia at the base of the prostate which separated the perineum from the pelvis and which provided a barrier to urinary infiltration to the rectal region. By careful dissection, he then discovered that such a fascia did indeed exist, and he named it 'the fascia of the prostate gland'. He then drew attention to the surgical implication of this fascia: that by performing the lithotomy incision above the base of the prostate, the fascia remained intact and provided the required barrier to the deadly urinal infiltration. He continued his lengthy article by describing in detail his recommended procedure for performing the operation. In the last paragraph he conceded that Mr Collies (sic) of Dublin might have seen this fascia but was unaware of its surgical importance; Pattison's only unequivocal claim was to have provided a rational explanation for the reduced mortality resulting from his modification of surgical technique.

Gibson read the article and, during a lecture to the medical students with Pattison in attendance, accused him of plagiarism in claiming Colles's fascia as his own discovery and of making unfounded and untrue assertions regarding the surgical importance of the fascia. Astonished, Pattison arranged a lecture the following evening to refute the accusations. He sent a note to Gibson inviting him to attend, but the invitation was ignored. Pattison duly gave his lecture in Gibson's absence, and vigorously defended his claims and assertions.

Gibson then mounted a second attack by submitting to the *American Medical Recorder*, under the pseudonym 'W', a critical review of Pattison's original article. The accusations of plagiarism and of overstating the surgical significance of the fascia were repeated.

Next came a flurry of charges and countercharges in Philadelphia's

William Gibson, professor of surgery at the University of Pennsylvania.
(Malloch Rare Book Room, New York Academy of Medicine)

leading newspaper, Poulson's *American Daily Advertiser*. On Friday 3 March 1820, there appeared a reprint of a 'puff' for the *American Medical Recorder*, originally printed in the *Norfolk Beacon*. This was not written by Pattison, but it did contain this single significant sentence: 'The initial article [for the year 1820], written by Mr. G. S. Pattison, surgeon, is without doubt the most important paper on the subject of lithotomy that has appeared in any country for a considerable time past.'

The following day a letter written by Gibson under the pseudonym 'Aristides' was published in Poulson's paper. It charged that Pattison's original lithotomy article contained neither a new discovery nor an important practical precept. Monday, 6 March brought a letter signed by X,[13] later identified as Dr John Eberle, the editor of the *American Medical Recorder*. He supported Pattison, with the injunction: 'Let Mr. Pattison's arguments be fairly and openly refuted, or let them not be assailed by the ambiguous assertions of anonymous writers.' Finally, two days later, 'Aristides' offered further derisive comments on Pattison's paper.

This preliminary skirmishing led to another exchange of pamphlets. These, 103 pages in all, were replete with sarcasm and insults. This time it was Pattison who made the first move by responding in July 1820 to Gibson's attacks in *A Reply to Certain Oral and Written Criticisms, delivered against an Essay on Lithotomy, published in the January number of the American Medical Recorder*. After reviewing the events leading up to the dispute, Pattison turned his attention to the review by 'W' (Gibson). He found that it contained only seven pages, and that all but forty-three lines of these were quotations. He commented that the review reminded him of the story of an old professor who apologized to his class for not giving them a valedictory lecture by saying: 'I intended to have written you a very fine lecture, but, truth to tell, I am so morally and physically exhausted that I have found it impossible to compose one.' Pattison suggested that, in the same vein, Gibson had intended to write a spirited attack, but was so morally exhausted in composing forty-three original lines and so physically fatigued by copying the six pages of quoted passages that his 'amiable intentions were frustrated'.

Gibson responded a few weeks later with his *Strictures on 'Mr.*

Pattison's Reply to Certain Oral and Written Criticisms' in which he gave his version of the various events in the quarrel and refuted many of Pattison's assertions.

Pattison published the third and last pamphlet of the series in November 1820, after his move to Baltimore: *An Answer to a Pamphlet entitled 'Strictures on Mr. Pattison's reply to certain Oral and Written Criticisms, by W. Gibson, M.D.'* As well as responding to Gibson's two main charges, he dwelt at length on Gibson's accusations that he had been the aggressor in the quarrel and, further, that he enjoyed no reputation in his profession. The arguments in this and the two previous pamphlets were long and repetitive. Both antagonists were guilty of pomposity and pious self-aggrandisement.

The Gibson quarrel provided ammunition for Chapman during his own battle with Pattison. He declared that the alleged discovery of the prostatic fascia was 'a piece of imposture originating in falsehood, advanced with effrontery, and finally abandoned in disgrace'.[14]

It is clear that, although Pattison apparently believed he had discovered a new structure, he was willing to concede the prior claim of Abraham Colles when it was brought to his attention. Duplication of discoveries was common; libraries were far less extensive, and modern facilities such as microfilms and photocopying were 150 years away. Perhaps Pattison really had known about Colles's work before his arrival in Philadelphia and had hoped to hoodwink his American colleagues, but as such a ruse would be certain to be uncovered eventually, it seems unlikely.

Because urinary calculi are never removed now by the perineal-prostatic route, Pattison's paper is of historic interest only. At that time his recommended surgical approach at least prevented subcutaneous extravasation of urine by leaving Colles's fascia intact, and also avoided the risk of perforating the peritoneal reflection at the bladder neck (with almost inevitable death from peritonitis).

Peace finally prevailed. Gibson wrote a letter on 4 November 1820 to a mutual medical friend in Baltimore: 'My dispute with Mr. Pattison has, I hope, terminated. I regret very much the occasion of it, and the terms in which I have sometimes been obliged to speak of that gentleman. I have a respect for his talents, and hope sincerely that the school to which I am indebted for my present situation will be benefitted by his appointment.'[15]

Would that Pattison and his colleagues had heeded the wise observations of Sir John Barclay, the great early-nineteenth-century anatomist! His remarks are very apt:

> Gentlemen, while carrying on your work in the dissecting-room, beware of making anatomical discoveries, and above all beware of rushing with them into print. Our precursors have left us little to discover. You may perhaps fall in with a trifling supernumerary muscle or a tendon, a slight branchlet of an artery, or perhaps a minute stray twig of a nerve—that will be all. But beware! Publish the fact, and the chances are ten to one that you have been forestalled long ago. Anatomy may be likened to a harvest field. First come the reapers who, entering on untrodden ground, cut down great stores of corn from all sides of them. These were the earliest anatomists of modern Europe, such as Vesalius, Fallopius, Malpighi and Harvey. Then come the gleaners, all gather up ears enough from the bare ridges to make a few loaves of bread. Such were the anatomists of the last century—Winslow, Vicq d'Azyr, Camper, Hunter and the two Monros. Last of all come the geese, who still contrive to pick up a few grains scattered here and there among the stubble, and waddle home in the evening, poor things, cackling with joy because of their success. Gentlemen, we are the geese.[16]

Living and Teaching in Philadelphia

Because his professional relationships were so stormy and contentious, Pattison no doubt appreciated the comfort and support provided by his brother and sister near whom he lived on Walnut Street. Little information is available about his social life. He had tried, before leaving Scotland, to obtain some fashionable introductions, but his dubious background provided an impediment to social success. A Glasgow friend living in London had written a letter of introduction to his cousin, Dr Robert Hare, the newly appointed professor of chemistry at the University of Pennsylvania. Pattison, after hearing nothing, wrote to Hare the following year and received a cold response: 'November 17, 1820. Dear Sir, I have just received your letter of the 15th inst. That of my cousin Mr. Stirling of London to which you allude I did not preserve. It was such as to induce and justify the civilities shown to a respectable foreigner. I am, [Dr. *crossed out*] Sir, Your obt. servant, Robt. Hare.'

We find the Pattison brothers under happier circumstances at the Philadelphia Burns Club in January 1820.[17] John was in the chair and his brother was designated as 'extra croupier'. An illustrious gathering of Scottish expatriates enjoyed the excellent food, wine, music, readings, and toasts. John had composed some sentimental verses for the occasion, which he recited. ('It was sweet to repose mid the red flowering heather . . . Pure honour, and virtue, dear Scotland shall nourish . . . Our thistles shall bloom on our mountains and valleys . . .'). Pattison's address was reported in full in the *American Daily Advertiser* of 2 February. He tactfully drew an elegant parallel between the 'American Philosopher' (Franklin) and the 'Scottish Bard', concluding, 'I would say that our Franklin has gone to be a discoverer in another, a happier region; and that our Burns is now a poet in a land . . . where his poesy can only be equalled, and where his divine genius is unshackled from the fetters of penury, dependence and mortality.' The speech was received with 'rapturous shouts, and the immortal memory of the great Franklin was drank from flowing glasses'. Towards the end of the evening, Pattison was toasted by the assembled company, after which he was elected vice-president for the following year. Clearly, he was in his element and all the petty bickerings with Chapman and Gibson were temporarily forgotten.

On at least one occasion Pattison sought diversion outside the city, where he was observed by a local resident: 'Pattison I remember to have seen . . . on a mineralogical excursion in my neighbourhood [Frenchtown, N.J.]. I judge that the cause of his devotion to natural history is the cause of the want of something to amuse and diversify the *tedium vitae* of a bachelor's life.'[18]

A faint echo of his earlier troubles reached Pattison in October

1819 when, on reading the recently arrived *Glasgow Herald* of 27 August, he espied a large advertisement with an alarming heading:

SEDUCTION, DIVORCE, CALUMNY
NO NOVEL
Printed from attested documents
in vindication of one husband,
and for the instruction of all.

This referred to a book which Andrew Ure had completed, giving his full account of the divorce action with Pattison named as the transgressor. Pattison immediately sent a denial to the *Glasgow Herald*. The book was suppressed only days before publication when Mrs Ure confessed to her duplicity regarding Ure's lewd letter. The book relied heavily on her perjured affidavit; by its retraction Ure would have been liable to criminal prosecution had the book been circulated. Considerable expense had no doubt been incurred in the preparation and printing of two thousand copies; Pattison suggested sarcastically that, in order to recoup some of his money, Ure might now 'allow his *friend* Dr. Chapman to have the whole edition upon most moderate terms'.[19]

Pattison's in-fighting with his medical colleagues did not prevent his giving lectures, a procedure which in those days was both formal and dignified. Professors lectured in full dress, including a dress coat with a high, stiff collar. (Gibson made a startling departure from this custom when he brought back from England a large selection of broadly striped waistcoats with trousers to match; these he wore at successive lectures to prolonged rounds of applause from the students.)

Having no university appointment, Pattison gave no formal lectures to the medical students during the year 1819/20, but his five-month popular course of lectures on anatomy and physiology was enthusiastically received. It was delivered two evenings a week for a course fee of $10. He proclaimed in the syllabus: 'We propose to examine the structure and organization of man; to contemplate and admire the wondrous phenomena which regulate his economy; and finally, by the exhibition of some of the inferior structures, to demonstrate that He who created a worm, required to be the creator of a universe.' The course content was systematically outlined under the general headings of Systems (osteology, myology, neurology, and the circulation), Natural Functions (respiration, secretions, nutrition, and generation), and Sensations (the special senses). Some of the less orthodox topics included facial angles as determining the intellectual powers of the animal; muscles of the face as depicted in paintings; the windpipe and voice in song, stammering, dumbness, and ventriloquism; and the nature of mind or soul as connected with the physiology of the brain.

Having failed to secure the university chair, Pattison joined actively in crusading for a new medical school to rival the one which had rejected him. He learned, for example, that a student, John Galloway Whilldin, had included in his graduation thesis some passages and comments which Chapman and a few of the other professors considered objectionable. They therefore insisted that the passages be expunged before Whilldin could graduate. Pattison used the event to attack the University of Pennsylvania, and Chapman in particular, by publishing the whole story of the Whilldin thesis. It was, in fact, a peripheral issue emanating from the mainstream of the Chapman-Pattison feud, presented as an academic affront.

Whilldin had completed all the requisites for graduation including the submission of a thesis entitled 'On the Nature and Treatment of that State of Disorder generally called Dropsy'.[20] At his oral examination on 20 March 1820, Chapman expressed his approval of the pathological and practical views but objected to some passages in which Whilldin strongly denounced nosology. (This involved the classification of diseases into myriads of supposed varieties and subgroups, each of which required some specific and individual mode of treatment; it led in turn to the now obsolete concept of treating one particular disease or symptom without consideration being given to other associated bodily changes or conditions.) Chapman felt that Whilldin's remarks constituted an attack on the university and on his own teaching. Whilldin countered Chapman's criticisms by stating that he had written without the slightest reference to any individual or school, but that he claimed the right to express freely his opinions. Chapman declared his satisfaction with the explanation, and Whilldin was thereupon notified by the dean that he had been approved by the faculty for graduation, unanimously and unconditionally.

Thinking this decision to be final, Whilldin left town for two days. On his return he was astonished to learn that the medical faculty required some alterations to his thesis. Meeting with the dean, Whilldin offered to defend his views, but was commanded peremptorily to remove a portion of his thesis, with the threat that, if he refused, he would forfeit his degree. Whilldin agreed to conform so long as he would not be considered to have abandoned his opinions. It was Whilldin's offending medical diatribe that Pattison quoted in full, with editorial comment, in his article.

The expunged passages contained a spirited attack, written with youthful exuberance, against the 'monster nosology'. Attention was drawn to the 'mortifying fate of every practitioner who addresses his remedies to a name given to a variable combination of symptoms, instead of watching attentively their many changes, and varying his treatment accordingly'. Whilldin also eulogized Dr Benjamin Rush, whose 'gigantic powers shattered the fetters of nosological tyranny'. All of this must indeed have sat ill with Chapman, who himself taught the nosological approach in his lectures and who was an avowed opponent of Rush.

Pattison proceeded to defend the expunged passages with irony and sarcasm. He upbraided the faculty for stifling the promulgation of new or unconventional ideas. Gradually he led into a severe denunciation of Chapman and his colleagues for their practice of holding 'prep' sessions. These were regular classes of fifty or sixty students who—for a $100 fee—were quizzed for forthcoming examinations. 'It must be self-evident that from one year's quizzing, the most trifling and deficient student will acquire . . . a parrot-like habit of giving answers to questions; concerning the spirit, the science, and the philosophy of which, he remains in the darkest and most disgraceful ignorance.'

The intensity of feeling engendered by the Whilldin episode was surprising. It undoubtedly assisted those who, like Pattison, were pressing for a second medical school in Philadelphia.

Pattison's year in Philadelphia was far from happy. He had been disappointed by his rejection for the chair of anatomy; he had been continually involved in quarrels, particularly with Chapman and Gibson; and he had been subjected to professional and social rebuffs. All this changed when he moved to Baltimore in August 1820 to occupy the chair of surgery at the University of Maryland.

The University of Maryland

Pattison's move was the prelude to one of the happiest, most successful and most productive periods of his life. His achievements over the next six years were enthusiastically acknowledged by students and public alike. His teaching resulted in a fivefold increase in the number of his students, while his success in establishing one of the first modern residential teaching hospitals in the United States added further lustre to his reputation. He was remembered and quoted for many years after his departure. His years in Baltimore were the zenith of his life and career.

In 1820 Pattison was a handsome young man in his late twenties, displaying more than his share of vanity and arrogance. A few years later his portrait was painted by Chester Harding, a fashionable American artist who also produced life portraits of Presidents Madison, Monroe, and Adams. It showed Pattison elegantly attired, with dark eyes and a receding hairline.[21]

Ever ready to take offence if provoked, he was nevertheless generous, fair, and supportive towards his friends and colleagues. For example, in a letter to the publishers Carey and Lea, he refused the offer of a free book on geography and offered to purchase it on the regular terms: 'If the gentlemen of America who are placed at the head of the literary and scientific establishments of the country do not patronize native publications of merit, they in my opinion neglect one of their most important duties.' And although not one to parade his religious feelings, he stated in 1820 that 'no man can have a more firm and decided belief in the great doctrines of Christianity, nor a more profound respect for its genuine votaries than myself'.[22] Professionally ambitious, he demonstrated an interest and curiosity in such new developments as the medical application of the recently discovered galvanism,[23] and announced lofty plans, later aban-

Granville Sharp Pattison, by Chester Harding, 1826. From the catalogue of
the 1868 Glasgow Exhibition of Portraits. (Mitchell Library, Glasgow)

doned, for writing a series of volumes which would have comprised the surgical anatomy of the entire body.

The offer of the chair of surgery at the University of Maryland, was initiated by Nathaniel Potter, professor of the theory and practice of medicine, who, on his own responsibility, invited Pattison to Baltimore. In 1821 Pattison changed chairs to become professor of anatomy. He remained on the faculty from August 1820 to the spring of 1826.

The College of Medicine in Maryland, granted a charter by the legislature in 1808, became a constituent part of the University of Maryland in 1812. For the first few years the faculty was short of funds, facilities, and the goodwill of the populace. By 1812 the faculty realized that the future had to be faced with courage, and on their own credit and responsibility, engaged an architect for a new medical college. Thus was born the oldest continuously used medical school building in the United States. The major influence came from John B. Davidge who had lectured variously in surgery, anatomy, and midwifery, and after whom the building is now named. It contains offices, laboratories, and two beautifully pro-portioned amphitheatres, each capable of seating 500 students. But there is also a variety of small, irregularly shaped rooms, storage areas, unobtrusive spiral staircases, and concealed doors and trap-doors, which reflect an era when cadavers were stored in whisky barrels or other hiding places, and escape routes were essential for medical students threatened by mob violence.

Pattison brought with him Allan Burns's museum of anatomical specimens, which had been housed briefly at the University of Pennsylvania. His purchase of it from Andrew Russel proved to have been a successful investment: it brought him not only great prestige but also the handsome sum of $7,800 which was realized when he sold it to the University of Maryland in 1820. It contained over 1000 speci-mens of normal and diseased organs, and was one of the best in North America, if not the world, at that time. In spite of the large purchase price, the university considered it a bargain. Soon after his arrival Pattison persuaded his colleagues to spend a further $8,000 for a new building to house it; the rather plain brick building, originally known as 'the museum' and later Practice Hall, has since been demolished.

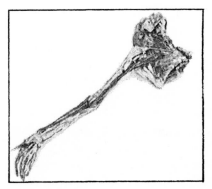

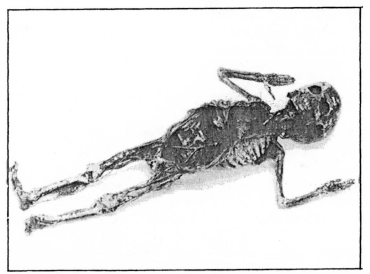

Three original specimens from the Allan Burns Museum of anatomical and pathological exhibits. (University of Maryland)
The museum has been a feature of the medical school at Baltimore for many generations, and in its early days it constituted the chief attraction for visitors to the university. The care with which the original specimens have been treated has varied over the years. Cordell in 1890 stated that the museum was in 'a lamentable condition'; Miller in 1919 made a similar observation. But about 100 of the original specimens have survived the intervening 170 years and are now well cared for by their present custodian.[24] The specimens, beautifully dissected and preserved, constitute a tangible link with Pattison and his career. Now, after playing their part in the Philadelphia quarrels, they lie in the sterile sanctity of a huge basement freezer.

The chief method of financing Pattison's ambitious schemes of expansion was by lotteries, which were well publicized and successful. Tickets cost eight dollars and prizes ranged from ten dollars to $30,000. The total prize money equalled the ticket sales; the profit to the institution came from a fifteen percent deduction from each prize. The managers of these lotteries included Pattison and most of his colleagues at various times.

General appeals to the citizens of Baltimore were made from time to time. On 15 May 1821, for instance, Pattison was a signatory to a letter in the *Federal Gazette and Baltimore Daily Advertiser*, in which notice was given that every city resident would be solicited for a contribution and for books for the university library. The letter noted that the medical school alone was bringing in $200,000 in business for the city every year.

Medical Education and Practice in Baltimore

The medical students at Baltimore usually lived close to the college in boarding-houses, at which room and meals cost three to four dollars a week. They frequently organized study groups, and seem to have been serious and well-behaved, but perhaps lacking the spontaneity of thought and action of their modern-day counterparts.

George Callcott, in his *History of the University of Maryland*, has provided a vivid account of the day-to-day life at the college, drawn from contemporary newspaper accounts, student notebooks, and other primary sources. Most students were eighteen or nineteen years of age when they matriculated, armed with some knowledge of Latin and Greek, and perhaps the experience of an apprenticeship with a practising physician. Usually two years of classes, each lasting about four and a half months, were necessary to receive the degree of M.D.

On arrival in Baltimore in October, the new student must have been appalled to hear the screams of patients undergoing surgery without anaesthesia, and to see and smell corpses in various stages of dissection. He paid a five dollar matriculation fee to the dean and twenty dollars to each of his seven professors for tickets of admission to lectures and laboratory sessions encompassing anatomy, theory and practice of medicine, chemistry, materia medica, surgery, 'medical institutes', and obstetrics. The first week of term was taken up by introductory orations by each of the professors, after which the serious work began.

The two courses which might seem strange to today's students were medical institutes and theory and practice of medicine. The former, a much disliked course, encompassed a curious mixture of physiology, pathology and diagnosis, including primordial aspects of clinical chemistry. The course on theory and practice of medicine was the forerunner of internal medicine, dwelling on symptomatology and treatment of diseases. Nathaniel Potter delivered the lectures for nearly forty years from the same crumbling yellow lecture notes. Topics ranged from miasmic fevers to haemorrhoids, from dropsy to nymphomania. (This last condition was apparently considered to include satyriasis; the symptomatic cure was a diet of bread and water, cold baths, a hard bed and massive doses of calomel, but the permanent cure was 'matrimony'.)

The most popular course was anatomy, particularly when it was combined with surgery. Anatomy was held in high esteem because of the ready availability of cadavers, at $2 to $4 each, which the students were free to dissect at any time of the day. Every state had its own particular laws regarding grave-robbery. In Maryland it was technically illegal, but the punishment for those caught was merely a small fine. Hence the supply at the university was such that it was able to sell its surplus to colleges as far away as Maine, where, in the 1820s, the penalties were severe, including imprisonment and public whipping not to exceed thirty-nine stripes. In Baltimore and Philadelphia, corpses were almost always obtained from potter's fields (burial grounds of the very poor) where the dead were given a token interment before being uplifted by the universities' agents. Yet public abhorrence of anatomy was just as powerful in the United States as in Britain. It was Pattison's view that each student of anatomy and surgery should dissect six bodies a year in each of his two years at a university; only about half of this goal was realized when he was at the University of Maryland.

Medical knowledge was still in its infancy; practitioners were hampered by a total lack of the most basic modern-day facilities, such as stethoscopes, microscopes, thermometers, hypodermic needles, antibiotics, antiseptics, and anaesthetics. The students must have realized the inadequacy of the available methods of treatment. It was said that every good doctor carried a lancet in one hand and calomel in the other.

Bleeding patients of a pint or two of blood, taken by nicking a vein in the arm or neck with a lancet, often reduced fever, pain, and tension, and promoted sleep. An alternative form of bloodletting, suitable for localized pain, was cupping: for this, the physician made several superficial cuts in the skin with a lancet or a scarificator and then firmly applied a heated, thick-walled glass suction cup which was gradually allowed to cool. Yet another variant, leeching, was widely used for haemorrhoids and inflammations: the leeches sucked about an ounce of blood and then dropped off; they were dipped in vinegar to make them disgorge the blood and become available for immediate re-use. A procedure used to draw out poison was blistering, induced by an application of cantharides or by a coating of gunpowder which was touched with a lighted match.

Along with these brutally heroic measures went the universal use of the ubiquitous calomel. This toxic mercury salt was viewed by the physician as a purgative but felt by the patient as purgatory. It resulted in massive diarrhoea, often accompanied by rectal bleeding and associated symptoms such as sweating and salivation. At Pattison's time, patients and some physicians were starting to revolt against this time-honoured panacea. A poem which epitomized the feelings of the public appeared in a Virginia newspaper in 1825:

> Physicians of the highest rank
> To pay their bills we need a bank;
> Nor talents bright, nor art, nor skill,
> Preserve us safe from Calomel.
>
> Howe'er their patients do complain
> Of head, or heart, or nerve, or vein,
> Of fever, thirst, or temper fell,
> The medicine still is Calomel.

And when I do resign my breath,
Pray let me die a natural death,
And bid you all a long farewell,
Without one dose of Calomel.[25]

Most sick people were treated at home, where they were tended by their family and attended by a private physician or surgeon. Hospitals at that time were little better than almshouses. Supported by the wealthy and staffed by pious volunteers, they catered for the sick, impoverished patient. The worst were simply pesthouses where the indigent went to die.

In contrast to this primitive nature of hospital care, medical lectures encompassing the current state of the art were presented with flourish and pomp. Pattison must have given hundreds of lectures in the upper of the two amphitheatres at the University of Maryland; here too he did his public dissections. He found out early in his first year that he had to stand on one particular spot to be heard distinctly by all of his audience. There was inevitably some restlessness on the part of the senior members of his class, as each student had to attend the same lecture course consecutively for two years. Among his students in the 1821/22 anatomy class were his brother Frederick Hope and his nephew John.

Teaching occupied most of Pattison's time, and into this he threw himself with his customary energy and self-confidence. Students regularly applauded his lectures, and most of them assumed, as did he, that they had heard the world's greatest authority on whatever subject he discussed.[26] Within a year of his arrival he was elected dean. He gave one formal lecture a day. Each was carefully prepared, memorized, and presented with oratorical polish. Some measure of his success as a teacher can be taken from the fact that his students in anatomy increased from seventy to 347 during his years in Baltimore. (When boasting in 1822 about the size of his class to Valentine Mott, then professor of surgery at Columbia University Medical School in New York, Pattison quipped, 'I think we may by exertion sink Philadelphia *between* us!') The lectures occupied only part of his day. In addition, he performed operations, made rounds of the infirmary, carried out his share of administrative work,

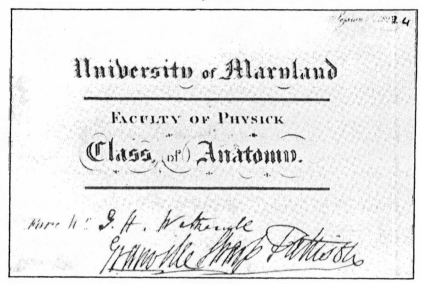

Ticket of Admission to Pattison's anatomy class in the 1823/24 session at
the University of Maryland. (University of Maryland)

contributed to the surgical literature, and waged his pamphlet wars
with Chapman and Gibson.

Pattison's skill and ability as a teacher constitute one of his main
contributions to anatomy and medical science. His forte was his
practical approach, one designed to produce competent, responsible
physicians and surgeons. His students were stimulated during his
anatomy lectures by his frequent allusions to physiology, pathology,
surgery, and literature. An eyewitness account of one of his lectures
at the University of Maryland appeared in a newspaper article
written by a doctor from the rival city of Philadelphia: 'I was
conducted to the college, which exhibited the animated picture of
327 young gentlemen, anxiously waiting the hour of lecture. . . . My
ears were charmed by the eloquence of the lecturer, Professor
Pattison. The intricate parts of the neck, beneath the under jaw,
were the objects of demonstration. . . . An old physician remarked
to me that he thought Professor Pattison the only man in America
capable of teaching anatomy with full effect.'[27] Similar observations
were made much later by one of Pattison's colleagues: 'His lisp and

Scotch accent made [him] one of the most interesting lecturers of his day. Few men in any age or country ever enjoyed such widespread popularity as teachers of anatomy as this distinguished Scotchman. . . . It is no exaggeration to say that no anatomical teacher of his day, either in Europe or this country, enjoyed a higher reputation.'[28] Many years after leaving Baltimore he was still remembered with admiration and affection for his 'unrivalled ability as a teacher of anatomy . . . brilliant and extensively known talents as a lecturer . . . animating his colleagues with his own zeal'.[29]

One of Pattison's students during his first year at the University of Maryland, William Somervell, kept detailed, closely written notes, which show the solid, factual content of Pattison's lectures. Containing no *aides-mémoire*, they must have been difficult to study, but the tedium of dry anatomical detail was alleviated by frequent clinical illustrations. Pattison was still recounting his discovery of the prostatic fascia but giving Colles credit for first describing it. (In common with countless students, then and now, Somervell invariably referred to the 'prostrate' gland!)

At the end of their two years of study, students were required to present a thesis and pass a final examination. The thesis could be written in Latin or English and a gold medal was awarded for the best thesis in Latin. It cost $20 to take the final examination. 'The nervous young man was invited to sit at the head of a large table around which the entire faculty was assembled. Slowly at first, and then faster, the professors fired questions. After stammering answers for an hour, the lad was invited to leave the room.'[30] A majority vote of the faculty was necessary to pass; a tie qualified the student for a second examination.

Commencement came in April. On the night before the event, the faculty gave a banquet for the graduates. Next day a band led a procession to one of the amphitheatres. 'On Professor Davidge (the father of the institution) and Professor Pattison making their appearance, they were received with three distinct bursts of applause by the whole assembly.'[31] These were indeed happy and successful days for Pattison.

More Quarrels

While tremendous progress was apparent in new facilities and increased student enrolment, dissent was ever present among the faculty. Because the medical college was a private institution largely financed by the members of the faculty themselves, disputes could not be referred to any hierarchical authority such as a president. Pattison participated in two quarrels.

The first involved Professors Davidge (surgery) and Elisha DeButts (chemistry); they were renting off-campus rooms for evening sessions of coaching, and charging students $10 a term for their 'Medical and Chymical Conversations'. The students came to feel that attendance at these was an unlisted requirement, and Pattison and other faculty members felt that Davidge and DeButts were enriching themselves unfairly at the students' expense. Pattison had berated Chapman for this same offence a few years earlier. He was appointed chairman of a committee to resolve the dispute; its report was presented to the entire university faculty on 16 November 1824. By a vote of twelve to six with two abstentions, the regents (members of the governing board) decreed that no professor in the faculty of physick could receive any emolument from the students other than the regular price of their tickets of attendance. Pattison thus incurred the wrath of two of the original founders of the institution for the public humiliation and loss of revenue which they had suffered.

The second quarrel revolved around Horatio Gates Jameson, Pattison's senior by thirteen years. An able and ambitious surgeon, he and Pattison had been sniping at each other for years. As early as 1820 Jameson had sided with Gibson in attacking Pattison's paper on lithotomy; he pronounced (wrongly) in a medical journal that there was no such thing as a prostatic fascia and that Pattison's practical inferences were valueless. In 1822 Pattison published a paper on a surgical procedure for correcting an arterial defect in the neck by ligation. A few months later came a blistering critique over the signature 'H.J.' (probably Horatio Jameson) which attacked Pattison on several counts: his lack of background knowledge, his incompetent surgical technique, and his ignorance of anatomy. The last accusation was supported by the report that, during the operation, Pattison mistook the omohyoid muscle for the carotid artery, and was prevented from ligating the muscle only on the intervention of a friend. True or not, Pattison provided no rebuttal at the time. But the remarks must have rankled because many years later, in 1833, he publicly ridiculed the story of the omohyoid muscle before his students at Jefferson Medical College, Philadelphia. ('The malevolence [of such a tale] may make [me] smile, but most certainly it will not make [me] angry.')[32]

Yet another public wrangle took place in 1823. Two years earlier, Jameson had reported the successful surgical removal of an extensive tumour from the upper jaw of a patient. Subsequently, he received reports from his friends and colleagues that Pattison was denigrating the operation, remarking that 'the tumor was returning . . . the case will terminate in Jameson's disgrace . . . the operation was imperfectly performed . . . I was anxious to have obtained the case.' Jameson sent a letter to Pattison requesting that he avow or disavow that he had expressed these views. When Pattison failed to reply, Jameson published the exchange and included a certificate from the parents of the patient stating that their son was in perfect health two years after the operation.

The quarrels climaxed when Jameson applied for a faculty position at the University of Maryland. He had been given informal verbal promises of success by a few of the professors. Pattison and some other professors, having no use for Jameson, were incensed that they had not been consulted, and succeeded in blocking the appointment. Jameson was so angry that he proceeded to establish a rival medical school in Baltimore, the Washington Medical College of Baltimore, which survived for a few years. This led the legislature to realize the folly of proprietary education and subsequently to assume control of the university under a board of trustees. Ironically, Pattison, by promoting Jameson's rejection, had brought about a political move which led in turn to his own resignation.

Little first-hand information exists regarding Pattison's social life in Baltimore. That his Philadelphia enemies failed to have him ostracized is certain; newspaper accounts of his activities are almost invariably favourable. It is likely that he attended the First Presbyterian

Church (his niece was baptized there in 1823). Yet one Baltimore gentleman remembered Pattison as being morally suspect. He was said to have taken so much mercury (for the treatment of syphilis) that he was afraid to take hold of doorbells for fear of an electric shock. Real or imagined affairs with ladies of fashion were still recounted in Baltimore sixty years after his resignation. Reports were circulated that he led such a gay life that he undermined his health to the point where his recovery was considered doubtful. Indeed, his ill health at the time of his departure was mentioned in the local newspapers and he himself referred to it in a private letter in 1826. Despite these reports, Pattison was to lead a full and vigorous life for a further twenty-five years.

The Pistol Duel with General Cadwalader

The comparatively even tenor of Pattison's life at this time was soon to be broken by another serious incident, a legacy of the Chapman quarrel. Again it involved a challenge; this time the challenge was accepted. For well over a century only the bare facts were known about this *affaire d'honneur* between Pattison and General Thomas Cadwalader. More recently, however, a bundle of letters has been found bearing the general's notation 'Affair with Professor Pattison'. These twelve numbered letters and interesting background information have been published by Nicholas Wainwright. They tell the whole story.

Cadwalader and Chapman were brothers-in-law, both married to daughters of Colonel Clement Biddle. Both were prominent in Philadelphia society. We have already seen that Chapman was a popular physician among the wealthier classes in Philadelphia. Cadwalader was a recognized leader of the city, a hero of the American wars against the British, a confidential agent and later chief director of the Bank of the United States, agent for the Penn family, and a leading literary figure in Philadelphia. He was a man of integrity and character who would brook no insolence or insult.

The general and Chapman had both served as managers of the long-established Philadelphia Assemblies; it was in this capacity that the general wrote a confidential letter to a friend and colleague on 23 March 1823 regarding the forthcoming annual ball. He was explaining a note which he had written ten days earlier. 'The object of the note was to prevent any individual manager from giving a ticket for the Assembly to Mr. P. I was desirous that the application, if made, should be made to me, as I was and am willing to assume all the responsibility of preventing Mr. P. from appearing at the Assembly, if practicable. Had any gentleman made the application to me in behalf of Mr. P. . . . I should have stated to him that after the quarrel between Dr. Chapman and Mr. P. and all the circumstances attending it, it would be unpleasant to a large number of Dr. C.'s friends and to myself especially, to meet Mr. P. in society.'

It was only a matter of days before Pattison learned of the note and he was not slow to respond, in a letter to Cadwalader of 26 March: 'Whilst in Philadelphia last week, I was informed that you had written a note to the managers of the Philadelphia Assemblies, directing them, if applied to for a ticket for me, to refuse it, *you being "willing to assume all the responsibility"* for preventing me from appearing at the ball. I now address you for the purpose of demanding either an explanation or satisfaction, for so daring an attempt to insult me. I have requested my friend to enclose you this letter and to adopt such measures as he may judge expedient for bringing this affair to an honourable termination.'

The letter was enclosed in a covering letter of the same date from his 'friend', who was, in fact, his brother Alexander Hope Pattison, then a captain in the British army: 'I have the honour to enclose you a letter from my friend Professor Pattison, and beg leave to observe that if you feel unwilling to offer an explanation perfectly satisfactory to his honour, I am ready to correspond with any gentleman whom you may name as your friend, and make arrangements to bring the business to a speedy conclusion.'

General Cadwalader wasted no time in replying (on 28 March) to Pattison's brother: 'Having no explanation to give in regard to the matter complained of by Professor Pattison, I have only to say that if he challenges me I will meet him, it being understood that the affair is to be considered as finally settled with its termination on the ground, whatever that termination may be; that the whole proceedings be conducted with caution and secrecy; and that any notice of the affair in the papers, hereafter, be avoided. It will be better to have the meeting on the border of Delaware and Maryland, as the reappearance of Mr. P. in this city or my visiting Baltimore might attract attention.'

Alexander replied to Cadwalader on 30 March, enclosing a formal challenge from Pattison: 'From the attempt you made to insult me whilst I was in Philadelphia, I request the satisfaction due from one gentleman to another. My friend who encloses this communication is fully authorized to make the necessary arrangements.'

On 31 March, Alexander served notice that he was handing over his charge to Jonathan Meredith, who the same day wrote to the general regarding detailed arrangements. By return mail the general replied to Meredith, acknowledging receipt of Alexander's and Pattison's letters. 'The latter contained a challenge which I accept.' The general named Captain A. J. Dallas of the navy as his friend and enclosed a letter from him:

> Sir, The enclosed letter from Genl. Cadwalader will shew you that I am authorized to act for him in the affair with Professor Pattison. In compliance with your letter, I will be at New Castle, at the Steam Boat Inn, on Friday evening [4 April], in the expectation of seeing you early on Saturday morning by which time you will have arrived there, leaving Baltimore on the boat of Friday. G.C. will accompany me, and will remain at a convenient point in the neighbourhood. It is expected that Mr. P. will be with you. The time is fixed to prevent inconvenience in case of the mail being late in arriving at Baltimore on Thursday. The preliminaries can require but a few minutes to adjust and, within a few miles of New Castle, a convenient place of meeting can be fixed upon, so as to afford easy access to either Pennsylvania or Maryland, after the affair is over.
>
> The following to be the arrangement *on the ground*.
> 1st. Pistols to be loaded by the seconds in the presence of each other.
> 2nd. Parties at ten paces distant and to remain in limb and body with as little motion as possible.

General Thomas Cadwalader, by Thomas Sully, 1816. Published by
permission of the owner, Captain John Cadwalader. (Historical Society of
Pennsylvania)

3rd. Word to be given, 'Gentlemen, are you ready? *Fire*.'

4th. Pistols pointed downwards and not to be raised until the word *fire* is uttered, after which the parties may fire when they please.

5th. No rest to be taken of any kind.

6th. The affair to be considered as wholly terminated by the meeting, whatever the result may be.

I beg you to bring with you a transcript of the foregoing arrangements, signed by you. The same shall be done on my part so that we may exchange them.

The last letter was Meredith's reply, written on the day of the duel, agreeing to the six conditions, which he wrote out again over his own signature.

The duel was fought on Saturday, 5 April 1823 near New Castle, Delaware, at the mouth of the Delaware River. This was a well chosen location, being roughly equidistant between Baltimore and Philadelphia; three other states (Maryland, New Jersey, and Pennsylvania) are all within a few miles of it. By the following day, rumours of the duel had reached Washington.

No description of the duel has been found. It is believed that the ground was in some marshes a few miles from New Castle and that John Davidge himself was the attending surgeon. The outcome was that Pattison was unhurt, 'but a ball passed through the skirt of his coat near the waist'.[33] Cadwalader, aged forty-three, was severely wounded: Pattison's shot entered his right arm near the wrist, traversed the length of the forearm and lodged in the elbow. The location of the ball was such that it could not be removed by surgery. Although Captain Dallas, the general's second, glibly remarked after the duel that the wound was only a scratch on the outer part of the arm, the general apparently did not go home for two weeks, to the great anxiety of his family. When he did return, he dictated all his letters and signed with the left hand. The general never recovered the full use of his right arm and the wound gradually undermined his health and hastened his death eighteen years later.

The *Columbian Observer* of 10 April 1823 provided an interesting comment on Cadwalader's behaviour: 'We understand there was no "*squatting in this business*", but merely a little bullying—as for instance—it is said Mr. C. told his antagonist that, *if* he had not been *wounded*, he, Pattison, would have been a *dead man*.'

Cadwalader's friends were quick to denounce the 'Scotch villain': 'I am writing to you to learn . . . if it be true that this reptile has obtained such an advantage over you. . . . For your sake, chiefly, I rejoice that his *life* was not terminated by your hand, though I should scarcely have regretted that, through your agency, he had been winged in *his* right arm. . . . [He] may be safely trusted to the certain reward which sooner or later never fails to reach animals of his nature.'[34] General Winfield Scott, a close friend of Cadwalader since the war of 1812, offered a more thoughtful comment: 'Several persons in my hearing have remarked on an apparent inconsistency on your part. But to me your conduct was just what I should have expected. When it was a question of admitting a *doubtful* character to an assembly of ladies you were naturally scrupulous; but when the affair took the *other turn, the soldier predominated.* Your conduct was perfectly natural; but I am one of those who believe P. not a gentleman.'[35]

Little is known about sentiments (if any) expressed on Pattison's behalf. 'Patriots and moralists alike were shocked, and besides blaming Pattison and foreigners generally, they also blamed the University [of Maryland] for harbouring such a man.'[36] Pattison occasionally made passing reference to the affair later during his years in London where he displayed the pistols on his drawing room table as a casual reminder of his victory. Pride, rather than shame or regret, emerges as his prevailing emotion.

The Baltimore Infirmary

A few months after the duel, Pattison's greatest contribution to medicine and medical education was made. This was the establishment of the Baltimore Infirmary, a pioneering step in modern clinical education in the United States. As late as 1849, only a quarter of medical schools reporting to the newly organized American Medical Association required students to attend a hospital at all.

Even in the 1850s clinical instruction was a major problem of nearly all medical schools in the United States. In Europe, a few countries had instituted some modern hospital teaching in the previous century; but the earliest of the true teaching hospitals in London, the Charing Cross Hospital, was founded only in 1821; its primary purpose was to train doctors rather than to care for the sick.

When Pattison arrived in the United States, classroom lectures and demonstrations were the standard fare in medical schools. These were favoured by many educators in reaction to the empirical type of instruction imparted to doctors' apprentices. Clinical education consequently became badly neglected. The most that a student could expect was a tour of the local hospitals when the professors were making their rounds. Rejecting the system as a halfway measure, Pattison persuaded the faculty to inaugurate a hospital with living-in accommodation for students, enabling them to provide round-the-clock service. Such an arrangement, unique at the time, distinguishes the infirmary as one of the first modern residential teaching hospitals in the United States.

The faculty chose a site across the street from the medical college. The Baltimore city council had originally voted in favour of financing it but were overruled by the mayor, who exercised his power of veto. The building was to cost $11,589, with an additional $2,520 for furniture and fittings. Even with a mortgage of $4,800, $9,309 still had to be raised. Since the university corporation's line of credit at the local banks was void, Pattison persuaded his colleagues to obtain a personal loan of $7,000 from the Bank of Baltimore on their own individual surety and to provide the balance from their private resources. The infirmary became the professors' private property and was in no way under the control of the regents of the university corporation. It was administered solely by the professors who, in addition to their regular teaching duties, provided their students with extensive bedside instruction at no extra cost.

The cornerstone was laid on 10 June 1823. Remarkably, the building was completed and ready to receive patients fourteen weeks later. It was a handsome four-storey structure in the Federalist style, with two curved staircases leading to the upper floors. There were four wards, one of which was reserved for eye cases; capacity

Baltimore Infirmary, built in 1823. From *A University Is Born* (Medical and Chirurgical Faculty of Maryland)

was estimated at 160 beds. The semi-circular operating theatre was in the rear, with tiers of wooden seats capable of holding several hundred students. Behind the building were wooden huts for laundries and the 'necessaries'.

Pattison had so far succeeded in spearheading the acquisition of a new infirmary and in promoting therein the concepts of the modern teaching hospital. His final, splendid stroke was to invoke the aid of Roman Catholic nuns in providing the nursing services. There were few trained nurses at this time. Nuns had previously served heroically in European hospitals during the Napoleonic wars and were beginning to work as nurses in the United States. It was to the Sisters of Charity from nearby Emmitsburg, Maryland, that Pattison turned.

He made his first approaches through Mrs Mary (Marianne) Patterson, a lady of immense wealth and social standing.[37] Father John DuBois, a French missionary who had come to the United States in 1798 and who at this time was at Mount St Mary's, Emmitsburg, replied to Mrs Patterson on 10 May 1822: 'I can hardly express the joy [our sisters] felt at the opening which your Charity offers to their zeal. It is truly now that they feel they are true Sisters of Charity.'

Father DuBois wrote to Mrs Patterson again on 9 June 1822 on a variety of administrative and religious matters: 'I foresee some difficulty in procuring [the sisters] the consolations of religion, particularly the frequentation of the sacraments and the divine sacrifice of mass. . . . I think we could prevail upon one of the priests of the seminary to go to the infirmary at least once a week, and I am sure the good sisters would make the sacrifice of the daily mass to serve the poor; but this could not be expected for Sundays. . . .'

Following this correspondence, a long quasi-legal Agreement was drafted in 1823, the seventeen articles of which made provision for the spiritual, medical, and temporal care of the patients and the sisters while ensuring the smooth operation of the infirmary. The sisters, 'being ready and willing to fulfil the most menial or disgusting offices for the sake of Him who did not disdain to annihilate himself for us poor sinners', were to have no full-time female assistants, but were empowered to hire as many male servants as they thought proper. And, 'whereas experience has proved that the constitution of most women in this country is frequently injured by much washing', the sisters were at liberty to hire coloured women by the day for laundry duty, 'the linen and clothes of the medical students excepted'. The patients' personal clothing was routinely removed and then washed, cleaned, and mended before discharge, because many of the sick, 'particularly the blacks, may have ragged dirty clothes, even with vermin, by which the bedclothings and even the very air of the wards might be infected to the great injury of their health'.

Article Twelve was unpopular, particularly with the students: 'After ten o'clock at night neither the inmates of the infirmary, officers, students, hired people or even new patients shall be permitted to go out or to come in the house without a special permission from the attending physician, or any other person appointed by the board. At ten o'clock, the keys shall be brought to the head sister, who will keep them in her room, and will see that no abuse should exist in this respect through the connivance of the porter.'

By the summer of 1823, Pattison felt that it was time to implement the agreed proposals with the sisters. On 11 October he informed Father DuBois that the infirmary would be completed and ready for the reception of patients by 20 October and requested that he arrange for the sisters' immediate transportation to Baltimore. Sister Servant Joanna Smith duly arrived as the first manager of the infirmary, soon to be joined by Sisters Ann, Adele, Rebecca, and Barbara.

To prevent misunderstandings, regulations governing patients were drawn up too: (1) each patient was required to pay in advance a weekly fee of three dollars to cover board, medications, nursing care, and medical and surgical treatment; (2) no patient was to leave the infirmary without permission; (3) the patients' diet was to be under the exclusive control of the attending physician or surgeon; (4) any disorderly or intoxicated patient could be expelled by the sister superior or senior student; (5) all patients in the event of casual absence were required to be in the infirmary before dark; (6) smoking was prohibited in the wards; and (7) liquor was prohibited anywhere in the infirmary.

Many of these rules and others were summarized in the *Federal Gazette and Baltimore Daily Advertiser* of 24 October, in the name of the board of managers of the infirmary and over the signatures of the president (the state governor) and the vice-president (the mayor of Baltimore). One of the managers was Jonathan Meredith, Pattison's second at the duel. Attention was called to the requirement of consultations before the performance of major surgery, to the fact that all medical and surgical advice and services were free, and to the regular audible reading of passages from the Bible every Sunday in each ward.

The duties of the resident students were likewise announced. They were required to be in attendance at all hours except meal times. No patient was to be admitted after 9.00 p.m. in the winter or 10.00 p.m. in the summer. All students were to accompany the professors on their rounds and to write the histories of new admissions. The senior students had many responsibilities: keeping books, attending all white female patients, assigning patients to other students, visiting all patients before retiring, checking all prescriptions, collecting patients' dues, and submitting a weekly report to the treasurer. The other students attended their assigned patients, filled all the prescriptions, and did the necessary dressings.

All resident students were required to pay $300 per annum for room and board.

Such a heavy workload started to cause trouble among the students who, like interns and residents today, felt oppressed. They were irritated by other matters too: frequent unannounced visitors invaded their privacy; the irregular attendance of certain professors raised doubts as to who had the final responsibility for patient care. To resolve these difficulties, the dean forbade visitors in the students' private rooms for any reason and ruled that charge of the patients rested with the attending physicians and surgeons, not the students.

Friction between the students and the sisters was soon evident. Frequent complaints were sent to Father DuBois regarding the students' behaviour and disobedience. In a letter to Pattison on 7 January 1824, Father DuBois discussed the 10.00 p.m. curfew: 'I was fully aware of the opposition [that] this *necessary* rule of discipline would meet with on their part. . . . I put off the hour of shutting the doors as late as I could . . . still, it appears that your young gentlemen are not satisfied. . . . The clue of the business is a spirit of independence which prevails among our American youths. They have little idea of the momentary slavery which they must submit to in order to become skillful in their profession. They want to unite the pleasures of life with the serious studies of the medical art—and spoil them both. Happy yet if at last, the brightest talents are not swallowed up in the vortex of dissipation.' Father DuBois then contemplated the effect that a later curfew would have on the sisters: 'What rest can [they] have after a long day of hardships if, after they have retired, they must be continually disturbed by the ringing of a bell or rapping at the door until perhaps two o'clock in the night?' And he objected to having students on call for only part of the night: 'If it is objected that their presence is necessary at night in the infirmary, why would it be more necessary from twelve o'clock at night until daybreak than from dusk until twelve o'clock?'

Father DuBois's next comment referred to the regulation assigning the infirmary key to the head sister during the night: 'If the head sister does not keep the key in her room, who will keep it? Will it be the young men themselves? Then every one of them must have a key, as none ought to have a privilege above the others. Will it be left with the porter? But remember that the porter is a kind of servant liable to be bribed, to be imposed upon, to be *terrified* into *compliance* or *connivance* if not by threats at least by every *abuse, mortification* and trick which youth and vice can invent to disgust a man of his situation.' He acknowledged that locking the doors at night made access to the outside 'necessaries' impossible, but suggested that an easy chair might be provided in some remote closet of the house.

Finally, Father DuBois pointed out that he had always been opposed to resident students: 'It appears to me that the $300 board they have to pay will be fully spent if not overrun by their expenses—that their services are by no means *necessary*—that the sisters can easily be brought to do everything the young men have to do—dressing wounds, bleeding, administering medicines, etc., except in cases where modesty would not permit it, and even then if really it was necessary to do it at night, the porter might be employed by them to do that which men alone could do with propriety.'

One must feel considerable sympathy for this elderly, querulous priest, as he wrestled with the never-ending problems of the 'generation gap'. His complaints were directed at improving the comfort and well-being of the sisters and their patients; he would have been happy to see the sisters take over completely.

The problems were ultimately resolved. The infirmary broke even under the efficient management of the nuns. However, there were frequent references in the minutes of later years to difficulties with the nursing service, and the eventual retirement of the Sisters of Charity in 1879 was inevitable.

Departure from Baltimore

A proliferation of new proprietary medical schools in Maryland caused the legislature in Annapolis to give careful consideration to alterations in the act of incorporation. Each new charter meant fewer students for the established schools, who responded by intensifying their recruitment campaigns. One ruse employed to lure new students was to lower standards, making it easier to graduate. Increasingly flamboyant appeals offended doctors and politicians alike. For these and other reasons, both houses of the legislature approved in March 1826 a bill giving the state authority to assume control of the University of Maryland under a board of twenty-one trustees, 'none of whom shall be professors, or have any personal interest to be affected'.[38]

During Pattison's time at the University of Maryland, all records, articles, and advertisements about the infirmary emphasized that it was the property of, and under the immediate control of, the medical professors. Now, with one legislative stroke, the infirmary became the property of the state, although the professors were still carrying on their free work there and the building had been financed by their own private funds and personal loans. The final blow came when the new trustees took possession of the infirmary, refused to reimburse the professors for their stake in it, withheld the income derived from it, and even debarred the professors from the building.

The professors, now lacking tenure or security, were notified that they must resign and reapply for their positions. They were refused representation at meetings of the new trustees. Furthermore, a son of one of the trustees was appointed supervisor of the infirmary. All this humiliation and loss of privilege must have infuriated Pattison. It was no doubt with some relief, then, that he departed for England in the spring of 1826, ostensibly on account of ill-health. Indeed he was ailing: 'I confess I feel very far from well,' he wrote from shipboard, 'but I hope when we are fairly out my feelings will be more agreeable. We have seven passengers who are all disposed to make me as comfortable as possible.'[39]

A few weeks later, in a letter from England, Pattison expressed to the trustees his concern that they were destroying the University of Maryland. Accordingly he tendered his resignation. This may have been the statutory resignation demanded of all the professors. The trustees begged that he return to Baltimore, but in vain: the executive committee of the trustees received his final resignation a year later.

Pattison had confided to no-one his plan not to return; his reasons seem to have been associated with ill health and ill usage. Little did he realize that new and worse troubles lay ahead in London.

V

The University of London in Revolt

1826–1832

WITH PATTISON'S ARRIVAL in Britain from Baltimore in the spring of 1826 began one of the most extraordinary and disastrous periods of his life. In Glasgow he had faced a major crisis during each of his three appointments; his problems had continued at the University of Pennsylvania with his feuds with Chapman and Gibson, and at the University of Maryland with the Cadwalader duel; true to form, it was not long after his arrival in London that he became the central figure in an imbroglio of great bitterness and intemperate vituperation. Many of the issues have a familiar ring.

Pattison's activities during the year following his departure from Baltimore are obscure, but apparently he kept out of trouble. It seems that he had pinned his hopes on being appointed to the chair of anatomy at the proposed new University of London, because as early as 1825 he was soliciting testimonials from American friends and colleagues. His list of supporters is impressive; it included the secretary of war, the secretary of the navy, the attorney-general, the secretary of the treasury, the secretary of state, and various religious and medical leaders, including Valentine Mott, then professor of surgery at Columbia University Medical School in New York. Another American who lent his support was Albert Gallatin, an earlier secretary of the treasury and a renowned statesman, diplomat, and financier, who was minister to Great Britain in the period

1826/27. Pattison wrote to Gallatin on 27 April 1827, recalling their acquaintanceship in Baltimore and requesting a letter of recommendation. Gallatin complied with the request for which Pattison thanked him in a note on 11 May.

In addition to this powerful American support, Pattison received favourable testimonials from British medical authorities, including Mr (later Sir) Astley Cooper, James Wardrop, surgeon to the king, and Sir James McGrigor, medical director general of the army, and from his cousin Lord Eldin (John Clerk, who had defended him at the resurrectionist trial). It is not surprising, therefore, that his appointment as the first professor of anatomy was formally confirmed by the council of the University of London in July 1827. (His election was carried by a majority of one vote.) The professorship of 'morbid anatomy' (the anatomy of diseased tissues) was added a few months later. As a founding professor of the university situated in the most influential city in the world, Pattison's future seemed bright.

The Anatomy Act

The first classes were scheduled to start in the autumn of 1828. During the preceding summer, Pattison testified as a witness before a committee set up by the House of Commons to examine the problems associated with the teaching of anatomy. As well as describing the various means (legal and illegal) of obtaining bodies for dissection, he emphasized repeatedly how distressing and undignified it was for young men of education (i.e., medical students), because of the strict laws governing the disposal of corpses, to be compelled to organize and practise illegal exhumation. Pattison's most significant contribution to the committee was his spirited argument for informing and demonstrating to the public at large the techniques and purposes of dissection.

When asked why he thought that dissection was so very greatly disliked and feared, Pattison replied that it was in part from its association with illegal exhumation, in part from its being stipulated in penal law that the bodies of murderers should be dissected, and in part from the general lack of knowledge and understanding regarding dissection. It was customary at the turn of the nineteenth century to make public profession of faith in immortality, and to insist that the human spirit outweighed in importance the earthly body that contained it. But it was felt too that if the dead body were disturbed, the future of the spirit would be placed in jeopardy. Pattison challenged these ingrained convictions. He believed that, if the mystery surrounding the work of anatomists could be dispelled, universal abhorrence would change to sympathetic acceptance, the public would be willing to consent to the dissection of deceased persons, and ultimately everyone would gain from the medical advances that would ensue from a more plentiful supply of bodies. His popular lectures on anatomy in Britain and the United States were part of his campaign to change the public view of anatomy.

Pattison included in his testimony some significant recommendations. He spoke strongly in favour of repealing the law that restricted dissection to the bodies of executed criminals; such a move would help to dispel the stigma of dissection. Following on from this proposal, he advocated that all bodies be allowed to become the property of surviving relatives, who might, if they wished, then dispose of them to reputable medical schools. Finally, he suggested that the bodies of strangers be treated in a similar manner by the parish officers. In support of these proposals, he recommended payment for each body, and full funeral rites on completion of the dissection.

With the authority derived from his appointment as professor of anatomy at the University of London, Pattison's testimony carried much weight. Indeed, his campaign over many years to improve the anatomist's lot must be considered as one of his most outstanding achievements in the fields of medicine and medical education. His endeavours undoubtedly contributed in no small measure to the eventual passage of the Anatomy Act in 1832.

Prior to its enactment, legal authority for the dissection of executed murderers had been laid down in *An Act for consolidating and amending the Statutes in England relative to Offences against the Person.* Covering fourteen pages, it dealt with all kinds of crime, including murder, poisoning, bigamy, rape, sodomy, abortion, abduction, assault, piracy, and other sundry offences, with punishment for most being death. Lesser crimes merited public or private whipping, imprisonment with or without hard labour, transportation beyond the seas, and fines. It was for murder alone that the law decreed that every body, after execution, be either dissected or hung in chains. After judgment had been passed, the convicted murderer was placed in solitary confinement, fed only bread and water, allowed no visitors, and finally executed two days after sentencing. If dissection had been ordered, the body was immediately conveyed by the sheriff to the designated surgeon or anatomist.

The Anatomy Act (*An Act for regulating Schools of Anatomy*) became law on 1 August 1832, with a minor amendment in 1871. Its provisions reflect many of the recommendations Pattison made to the select committee of 1828. The ultimate sentence of dissection for executed murderers was repealed; the bodies of murderers could thenceforth be either hung in chains or simply buried within the precincts of the prison. Subjects for dissection were to be legally available from anyone having lawful possession of the body of any deceased person. The main stipulation was the exclusion from dissection of any person who, in his lifetime, had expressed a desire that after death his body should not undergo anatomical examination. Moreover, a forty-eight-hour waiting period was mandatory following death, as was a valid death certificate signed by a physician, surgeon, or apothecary. Finally, it was required that the body be interred within six weeks following death. From these provisions, it soon became clear that there would never again be a shortage of cadavers for dissection and that grave-robbery had been eliminated as a profession: the age of the body-snatchers was over. The act even set forth a disclaimer stating that no licensed anatomist would be liable to any kind of prosecution for having a dead human body in his possession. Only eighteen years earlier in Scotland, Pattison had been subjected to prosecution and ill usage for this very offence. Indeed, by helping to eliminate the cause of a crime of which he himself had been accused, he had come full circle.

English Medical Education in the Early Nineteenth Century

The practice of medicine in England in the early 1800s has been described as 'an art founded on conjecture and improved by murder'.[1] Practitioners in those days were, by today's standards, very poorly qualified to deal with the health of the nation. Charles Newman, in his *Evolution of Medical Education in the Nineteenth Century*, has provided a valuable description of the prevailing practices.

At the time of Pattison's arrival in London the medical profession had reached the stage at which three groups of doctor were recognized: physicians, surgeons, and apothecaries.

Physicians comprised the oldest of the groups, commanding the greatest privilege and respect. They alone gloried in the title 'Doctor', and they alone boasted a university education. They were professionals, displaying the skill of learning, not of technology. They were never 'paid', but were 'reimbursed' for their expenses and accepted presents in gratitude for their service. The physician usually came from an upper-middle-class background, and the end product was a cultured and highly educated gentleman, well-versed in literature and the classics. An adequate knowledge of medicine was often of secondary importance; his art consisted of the treatment of diseases by drugs and the writing of complicated prescriptions.

Boys destined to become physicians usually attended a grammar school or, occasionally, one of the public schools. A university education frequently followed, often at Oxford or Cambridge, although the medical curriculum for the training of a physician was non-existent. For his own professional security and peace of mind, the student felt obliged to acquire some skill in pharmacotherapy. Such examinations as he had to pass were conducted in Latin; in the minds of the examiners, his ability in this often outweighed all other considerations. It was in postgraduate education, at home or abroad, that the sound physician obtained his technical proficiency. This he did by attending lectures, listening to his experienced colleagues, and 'walking the wards'. None of these activities was organized, and each individual selected his own means of achieving a balanced education. Many physicians were in practice after a mere one year of training following their general education.

Surgeons came from a social background similar to that of the physicians. It was in temperament that the surgeon differed, reflecting in part the terrible daily experiences that he was bound to witness in that pre-anaesthetic era. The objective of a surgeon's education was essentially practical; such general education as he might possess was a by-product of his earlier years, acquired in spite of (rather than because of) his surgical training. The technical desiderata were speed and dexterity, and for these an intimate knowledge of surgical anatomy was essential: in the 1820s 'surgical anatomy' was the study of the anatomical relations between organs and structures, and was the branch of anatomy which was so vital to the early surgeons. The surgeon, in short, was trained to be a skilful practitioner, not a learned one, and for him a university education was thought not to be necessary. In the minds of the public, he was of a lower order than the physician.

To practise surgery, all that was required were certificates of one course on anatomy and one on surgery, and to have attended the surgical practice in a hospital for one year; there was no need to have learned any medicine at all. Before acquiring these credentials, many potential surgeons underwent five years of apprenticeship, often to a general practitioner. This period was long and frequently extremely tedious, but it allowed the apprentice to learn the basic facts and philosophy of medical practice in a leisurely manner. His tasks were often menial: dispensing, seeing the poorer patients, delivering messages, keeping accounts, and writing up the daybook in Latin. Far from being paid for these services, the apprentice was required to pay a premium for the privilege.

Diagnosis formed a relatively small part of the teaching of surgery, for two reasons. In the first place, the art was so rudimentary that all known diagnoses were easy: what the surgeon could treat, he could see or easily feel. Second, surgery was concerned with external medicine, because the deeper parts of the body were inaccessible to the knife. The more difficult diagnoses were of little more than prognostic concern.

Whereas the proud physician often disregarded common diseases in favour of obscure conditions and diagnostic challenges, the surgeon undertook the treatment of any obvious complaint, whether surgically correctable or not. Venereal diseases, for example, were therefore viewed as the province of the surgeon. For these, salves and inunctions were applied liberally. It was only in treating complications such as urethral stricture and abscesses that the surgeon's manual dexterity was put to use. (Penile amputation among military personnel suffering from syphilis or gonorrhoea was widely practised by surgeons in the sixteenth century, but not in Pattison's day.)

Apothecaries were the lowest order of the three. Like surgeons, they did not require a university education. They derived largely from a lower socio-economic level, that of shopkeepers or even servants. Following the Apothecaries' Act of 1815, which raised the general standard of medical education throughout the country, apothecaries became what might reasonably be considered the forerunners of today's family practitioners. They treated patients to the level of their competence, and called in consultants when necessary.

The apothecary's education was wholly practical; the only Latin he was expected to know was that written by a physician in prescriptions. He was actually discouraged from being 'learned', lest he trespass in any way on the path of the more erudite and prestigious physician. Since he was fundamentally a shopkeeper, his training was naturally by apprenticeship, usually of five years' duration. When he was duly qualified, his duties were the attending of sick patients not requiring external or manual aid, and the prescription and supply of appropriate medicine.

All three classes of healer were taught along lines similar to those of today, with one significant exception: a detailed physical examination did not take place. The importance of taking a good medical history and making meticulous enquiry into symptoms was stressed; physical signs, however, were considered to be of far lesser importance. As the nineteenth century progressed, the value of physical examinations was gradually recognized, and with it a realization of the significance of physical signs. It became the practice for the discoverer's name to be associated with a sign—Argyll Robertson pupil, Babinski's sign, Hegar's sign, Kussmaul respiration, for example; this proliferation of eponyms, while affording the discoverer a modest niche in medical history, provoked groans from overburdened medical students. The introduction of physical signs into medical assessments, spurred by the discovery of the stethoscope, was to lead even later to newer techniques of clinical, laboratory, and radiological investigations. During Pattison's life in London most of these advancements were yet to come.

The principal method of teaching was by lecture, which provided a satisfactory introduction to the subject and, if skilfully delivered, stimulated the student to further reading and study. The lectures were at times enlivened by clinical lecture-demonstrations involving selected patients.

Teaching was conducted in various hospitals and in numerous private schools. In these establishments students had the opportunity to master techniques, but the staff were at liberty to do as much or as little teaching as they liked, using whatever methods they preferred. Much advantage was taken of the services of unpaid apprentices and paying pupils. The private schools flourished because they taught anatomy all year round, using an ample supply of bodies from illegal sources. The hospital schools, reluctant to make use of illegal suppliers, held anatomy lectures only in the winter, when the cold weather could preserve the bodies.

All students, even the most junior, had contact with patients from their very first day, but 'walking the hospital' was an inefficient and unprofitable exercise, amounting to little more than looking at the cases and hearing the remarks of the medical attendants. Symptoms were recorded in great detail, no physical examination was performed, and the usual solemn pronouncement for further treatment was 'Continuentur medicamenta'.

The medical history was never less than thorough. It might start with a meteorological report to trace the relationship of the weather to the progress of the disease, followed at length by the biography of the patient from infancy, and concluding with the symptoms of the existing illness, negative as well as positive, and of the condition of every function, whether disturbed or not. The only physical signs likely to be touched on were those associated with the facies, the tongue, and the pulse, while considerable attention was directed to the appearance of the blood (obtained by venesection), urine, and faeces.

Surgeons were more concerned than physicians with physical signs, because most surgically correctable conditions were external. Thus, the state of the skin over tumours and injuries had an important bearing on the outcome of operations; such findings were in reality a measure of infection, although surgeons were not then in a position to know this. Aneurysms, tumours, and hydrocoeles were all well understood and described; and the surgeon could tell if a patient had become infected after an operation.

One aspect of medical education as important then as it is now was the propagation of knowledge by example. It was the teacher's actual work and attitude toward his patients that counted for so much. The eminent medical men of the early nineteenth century were conspicuously lofty, but their prominence as great teachers brought the student into contact with first-rate methods of practice. This teaching by precept and example laid the foundation for the many outstanding advances that were so soon to come.

All serious students of medicine flocked to the continent, to France in particular, for postgraduate education. Here were the undisputed leaders and here, owing to more advanced legislation, was a plentiful and cheap supply of cadavers. The French school originally owed its pre-eminence to the need for surgeons for the Revolutionary Wars, and from this practical beginning came the great men of French medicine: Bichat, Laennec, and Magendie. Pattison benefited from visits to France from time to time and he espoused the work of the French anatomists by translating and editing books by Velpeau, Cruveilhier, and Masse during his later years in Philadelphia and New York.

At the heart of medical education of course, then as now, was the medical student. He was reported to be appalling: physically dirty, foul-mouthed, callous, and cynical. Smoking, drinking, and brawling were frequent occurrences in the dissecting room. Such behaviour was perhaps an outward and visible sign of a toughening process, a disciplining of the sensibilities, needed for the students' professional

This splendid caricature of a
19th-century British medical student
certainly confirms the contemporary
view of the 'dirty, foul-mouthed,
callous and cynical' student of
medicine!

development at a time when they were assailed by filth, squalor, and the general coarseness of society.

The teachers themselves, whose example the students were required to follow in the wards and dissecting rooms, were in some measure to blame for their students' behaviour and attitudes. The comment of one consultant to his students, that 'the sole object of your medical studies are two: first to get a name; secondly to get money',[2] is indicative of a certain cynicism among the profession. Dirty jokes and gross mnemonics abounded: 'The essence of the vileness of it was that elderly men talked in that way to boys just from school. . . . It was thought so funny, so valuable and so impossible to replace, that women students were objected to because they would spoil it all.'[3] (No woman attended medical school or practised in England until seven or eight years after Pattison's death.) During the ensuing decades, behaviour and attitudes improved in step with the dramatic advances in overall medical management.

The University of London

When its doors opened to students in 1828, the self-styled 'University of London' was without a royal charter, and thus was officially a private institution.[4] It was founded for the teaching of 'the youth of our middling rich people', who were unwilling or unable to attend Oxford or Cambridge, 'the two great public nuisances', as Jeremy Bentham called them; membership of the Church of England was necessary at both. The new University of London became a source of higher education for all noncomformists, Catholics, Jews, and those excluded from Oxford and Cambridge by social, financial, or educational restrictions. Moreover, its objective was to offer a broad range of courses, whereas the two older universities laid heavy emphasis on the classics and on the training of clergy for the Established Church.

The university was financed by the sale of £100 shares to a large number of 'proprietors' (shareholders), who in turn elected from among their number a twenty-four-man council. Later a warden was appointed who served as an emissary of the council. The proprietors, members of council, and the warden all played their part in Pattison's later troubles.

There was no senate, and the professors had absolutely no official status in regard to policy matters; this caused serious problems in the first few years of the university's existence. Another source of criticism was that the professors were elected on the basis of letters of recommendation and not by open competition. (In his application for the professorship of anatomy, Pattison had pressed for the election to be decided by public *concours*, in which he would have engaged with the other candidates. The proposal was declined.) The council naturally sought to gain men of high standing, but it could offer only small emoluments and a precarious future. The most conspicuous characteristic of the teaching staff was its youth: of the twenty-seven professors, twenty-one were under the age of forty, and six were under thirty. A high percentage were Scots.

The students came from all walks of life: there was no homogeneity here such as was enjoyed at Oxford and Cambridge. Nor were there any residential colleges, where some measure of moral and religious control could be exercised. The university provided a list of boarding-houses that had been inspected and approved subject to certain conditions, for example: 'that they will be bound to require their boarders to be at home at an early hour of the night; that they will not suffer gaming or any licentious conduct on the part of the boarders; and that they will require their boarders to be regular in their attendance at some place of public worship'.[5]

The students paid £28 for tuition for a full academic year. This sum was well beyond the means of the working class, being equal to the annual income of, say, a coachman. Pattison received £7 per student for his lectures, while his demonstrator received £5 for the anatomy demonstration course.

Full details of courses of study, attendance requirements, weekly revision examinations, and written examinations for the determination of winners of prizes and certificates of honour were outlined in 1831 in the first *London University Calendar*. It is significant that the general certificates required for medical practice refer only to attendance and not to proficiency or academic attainment; the folly of this system was roundly criticized at the time.

Toward the end of Pattison's years in London, the council, recognizing that there were in Britain some very good and some very bad practitioners, decided to offer as proof of competence a medical diploma, to be called 'The Diploma of Master of Medicine and Surgery in the University of London' (*M. Med. et Chir. U.L.*). It was required that each successful candidate obtain certificates of honour in all classes of medical instruction; attend a hospital for a year of medical practice and another year of surgical practice; prove his ability in Latin; present an 'essay' (thesis) that he had had to defend at a public meeting; and perform an anatomical demonstration. The diploma never achieved general acceptance because the lack of a royal charter denied degree-giving powers to the university.

The university year ran from early October to mid-May. From the viewpoint of an anatomist, the cold winter months offered the distinct advantage of retarding putrefaction of cadavers. The medical classes had a full and heavy schedule, starting at 9.00 a.m. (midwifery) and continuing with few breaks until 8.30 p.m. (medical jurisprudence). Pattison lectured every weekday from 2.00 to 3.00 p.m., with revision examinations at 1.30 p.m. once a week; at other times in the day he would be teaching in the dissecting room, preparing lectures, and participating in administrative matters. During his last year, much time was devoted to aggressive correspondence with the warden or the council.

The courses offered by Pattison in anatomy and surgery were described in the 1831 calendar in great detail.[6] The content of his lectures was essentially practical: descriptive anatomy (the study of individual bones, muscles, viscera, vessels, and nerves) and operative surgical techniques. This utilitarian approach differed from his earlier enthusiasm for general anatomy, introduced to England by Bichat and other French anatomists, in which an attempt was made to relate tissue composition of the organs of the body with their properties and functions. The latter system, which embraced physiology, pathology, and embryology, was seriously hampered by the total absence of histological and biochemical knowledge.

The dissecting room was open all day, with a demonstrator present at all times. Students paid £6 for cadavers, with the council defraying the difference between this sum and the usual charge of about £20. The Anatomy Act of 1832 had not yet become law; it is surprising that the official calendar of the university should publicize in print the fact that it was acquiring bodies by illegal means.

The students received extensive instruction in medical jurisprudence; the university calendar lists many subjects which would be treated today under paediatrics, genetics, obstetrics, and gynaecology. A few representative items are of particular interest: the legal privilege of not answering questions in the case of medical attendance at duels; the question of whether or not hermaphrodites exist; wagers respecting the sex of individuals; the right of children born in the absence of the husband beyond the period of uterine gestation; proofs of paternity; feigned diseases; imprisonment and execution postponed on account of illness; public health nuisances, such as drains and exposure to contagious diseases; the laws respecting military punishments and torture; the possibility of committing rape during the sleep or intoxication of the female; and so the list continues for six printed pages.

In a letter to the council in 1827, Pattison outlined his views on clinical teaching. He advocated that all medical professors be formally designated as clinical professors, and that they all teach clinical methods at the bedside as part of their duties. 'It would be in the power of the clinical professor to question the students at the bedsides of the patients, as to their own views of the nature of the diseases, and the reasons which have induced them to form them.'[7]

Examination questions for certificates of honour in the year 1829/30 are published in full in the 1831 calendar. The proof copy of Pattison's anatomy paper with his corrections and comments is in the University College archives. The twenty questions alone occupy two and a half printed pages. If answered in full, they would virtually have required the writing of a complete textbook, though the time allowed for the examination was only six hours. Marks for each question were indicated, the maximum total being 530. 'Describe the phenomena of ossification and state the period at which it occurs in the foetus, and the mode in which the phosphate of lime is deposited in the three classes of bones, the long, the flat and the short (20 marks).' 'Describe the anatomy of the urethra, the ducts which open into it, and the relations and connexions of its different parts (30 marks).' 'Describe the position of the ganglia of the nerves of organic life, situated in the head and neck, the branches they furnish and the connexions they establish (40 marks).' In a note at the end of the examination paper, Pattison added the comment: 'By proposing the queries in the manner which the professor has adopted, he has avoided all risk of putting leading questions, and he has afforded an opportunity to the distinguished student of showing the extent of his anatomical knowledge.'[8]

The students wrote a motto instead of their name on their answer sheets; this was, perhaps naively, intended to ensure anonymity. The motto was also written on a slip of paper, sealed, and finally opened by the warden at the actual prize-giving ceremonies, in order to identify the winners.

On the basis of the examination results, certificates of honour were awarded to a number of deserving students, and three winners were selected for the gold and two silver medals. All were distributed at a large, formal prize-giving ceremony each May. On these occasions, a speech was made by one of the professors. There is no record that Pattison was called on at any time for this particular function, but at times, he and his fellow professors read out answers that they considered to show exceptional merit. Occasionally, answers to questions were printed verbatim in *The Lancet*, to enable the medical profession to form an opinion of the attainments of the students of the newly opened university.

Storms of Dissension

Pattison's life in London began on a happy note. In April 1827, three months before his appointment at the university was announced and eighteen months before the start of his classes, he delivered a popular lecture on the heart and circulation to a large and prestigious audience at the Royal Institution. It was very well received and was praised in an editorial in *The Lancet*. Its significance in the fight for anatomical reform was hinted at in the last sentence of the review: 'We feel persuaded that these popular demonstrations will accomplish more in one year towards the removal of that hostility which the public have manifested against human dissections, than all that has been written on the subject during the last fifty years.'[9] Unfortunately, apart from this one favourable review, *The Lancet* was consistently antagonistic toward Pattison for the rest of his years in London, and indeed the editor, Thomas Wakley, later denied authorship of the review.

During the academic year of 1827/28, Pattison was requested by the university council to examine anatomical museums in Germany and to make a list of preparations for his classes. He was not due to receive any salary for his professorship until classes started in the autumn of 1828, but he agreed to undertake the trip on the understanding that he would receive a guaranteed salary of £300 per annum and travel expenses. He therefore abandoned the idea of returning to the United States for the intervening year, at what would have been a far higher salary. However, no minute was recorded by the council about this salary agreement; when he applied for it in December 1827, he was informed that the council declined payment, not feeling bound to honour their promise of the salary. This was an unfortunate beginning to Pattison's relationship with the university.

The next hint of trouble occurred in the spring of 1828, when he learned that the council had appointed a demonstrator in anatomy, James Richard Bennett, one of the unsuccessful candidates for the chair. It was the usual practice in those days for the professor to choose and appoint his own demonstrator, and Pattison was incensed that he had been denied this prerogative. He may have hoped to bring his nephew John from the United States as his demonstrator. Moreover, Bennett was to receive all of the £5 paid by each student for dissecting-room privileges; in all other London schools of medicine, the professor received both sets of fees and paid a pittance to his dependent demonstrator. Pattison thus found himself with a junior, wholly independent colleague 'whose interest it was to elevate himself by bringing the talents and the reputation of the Professor into disrepute with his students.'[10] George Birkbeck[11] had earlier warned the council about the likely consequences of such an appointment, but was overruled. Pattison must have realized by this time that frequent incongruous decisions by the council were to be expected.

During Pattison's three years of turmoil at the University of London, Bennett was never far from centre stage. Pattison's junior by some eight years, Bennett graduated with the degree of A.B. (Trinity College, Dublin) in 1817 and took the Letters Testimonial (i.e., the licentiate) of the Royal College of Surgeons in Ireland on 16 May 1820. He then moved to Paris, where he lectured to a class of English students. Encouraged by this experience, he proposed the establishment of a medical school there. But the Royal College of Surgeons in London and Mr Canning, the Foreign Secretary, conspired to block the plan, apparently because the College wished to see British students studying in England rather than in France. He then lectured in London for some years before applying for the chair of anatomy at the University of London. The announcement of Pattison's election and of Bennett's more junior appointment marks the beginning of their jealousy and rivalry.

It was not long after the start of classes in October 1828 that

Pattison noticed the first signs of trouble with Bennett. He had written a peremptory letter to Bennett, listing some muscles that he wished to be isolated for his next lecture and giving minute details about how the dissections were to be performed. Bennett in turn wrote, not to Pattison but directly to the council, proposing that the professor's dissections be relegated instead to two of 'the most clever pupils', who would receive some minor financial recognition and 'much distinction'. Two weeks later Pattison learned that some senior students were being admitted to Bennett's dissection class but were not at the same time attending Pattison's lectures. This reflected on his authority and prestige, while the loss of the sale of tickets to his lectures meant a reduction in his emoluments. On 22 November he wrote a letter of protest to the council. A compromise was reached, but problems of a similar nature were to recur in the ensuing years.

No further disagreements seem to have arisen until the end of the 1828/29 academic year, when, at the prize-giving ceremonies in May, Bennett allegedly behaved in a rude and offensive manner. On this occasion, Pattison read some passages written by the winner of the gold medal and expressed his admiration for the work. But Bennett exhibited contempt for the quoted passages. Edward Turner, professor of chemistry, reported: 'Bennett's conduct was extremely indecorous. His expressions of disapprobation were so loud, and his gestures so eager that I was quite alarmed and felt it necessary not to lose a moment in pacifying him.'[12] Bennett conveyed the impression to those around that Pattison was both ignorant and unable to detect certain anatomical blunders. The story of the incident was spread throughout the country, apparently in order to damage Pattison's reputation. Pattison claimed that he himself was unaware of the incident until several months after it had occurred, when he 'came to an immediate rupture with Mr. Bennett'. On 7 August 1829, he wrote a long, strong letter to the council, outlining his grievances.

Later the same month Pattison again wrote a letter to the council, this time complaining that Bennett was overstepping his authority by giving formal systematic lectures in anatomy; Bennett denied the complaint and outlined his interpretation of his duties as demonstrator. Word had spread about the dissension in the anatomy department, and two of Pattison's colleagues, Charles Bell, professor of physiology and surgery, and John Conolly, professor of the nature and treatment of diseases, offered to try to resolve the difficulties.[13] At the time of his appointment, Bennett had been granted 'chief direction of the dissection room'. It was now determined that Bennett was indeed in sole charge of the dissecting rooms, but that it was also his responsibility to prepare dissections for Pattison's lectures.

No sooner had the disagreement with Bennett been resolved than Pattison and Bell began to quarrel. During the summer of 1829, Bell and Leonard Horner, the influential warden of the university, had met in Edinburgh and had agreed between themselves—without giving the most distant hint to Pattison—that Bell would forthwith take over the chair of anatomy. This recommendation was communicated to the council, who promptly turned it down. Horner claimed later that it was Bennett who had suggested that Pattison and Bell change chairs. Bennett had written to him that 'the latter subject (surgery) is narrow and does not require much research or industry, and the anatomy, in Mr. Bell's hands, would be popular beyond measure'.[14] Pattison was convinced, however, that the original idea was suggested to Bennett by Bell.

Just before the academic year 1829/30 began, Bell wrote to the council complaining that Pattison was using the title 'Professor of Anatomy and Operative Surgery' on his class tickets. Since Bell was professor of physiology, surgery, and clinical surgery, there was clearly some overlap. Pattison responded in a testy letter ('I am really at a loss to comprehend what Mr. Bell will be at, or what will satisfy him.').[15] He pointed out that every anatomy teacher in London taught anatomy, physiology, and operative surgery, but that, to satisfy Bell, the council had 'lopped off physiology from my course'. Now it seemed that Bell wished operative surgery to go too. Moreover, Bell had approved the title of operative surgery on Pattison's tickets the previous year and had himself announced it publicly to the students. He therefore declined to alter his tickets for the balance of the academic year.

In December 1829 Bell again wrote to the council, complaining about another matter. Pattison had invited Baron Charles

Heurteloup, a French surgeon, to demonstrate to his class some newly developed instruments designed to facilitate the surgical procedure of lithotrity (breaking up bladder stones *in situ*). Bell felt that Pattison was again trespassing on his territory and refused to deliver any more lectures until he had received an explanation. Horner thereupon issued an injunction to prohibit Pattison from allowing Heurteloup to proceed with his demonstration. Not surprisingly, Pattison lodged a vehement protest against the injunction and the warden's interference: 'As I conceive I am entitled to teach . . . in the manner which I consider most for the interests of my pupils . . . and as I should hope that the council have sufficient confidence in the judgments of their professors to entrust to them the mode in which they are to perform their duties, I must most respectfully object to Mr. Horner's interference.'[16] Nevertheless, the council cancelled the baron's demonstration 'because of risk within the walls of the university'.

Far more ominous was the growing antagonism of the medical students themselves. In the previous academic year, two independent anonymous complaints had been lodged against Pattison. Both had given rise to full and independent enquiries, both of which had exonerated Pattison. He expected therefore that the council would refuse to recognize any further charges, but he was disappointed. Late in 1829 a new complaint was laid against him; this time his accusers identified themselves, but only on condition that their names be kept secret. The charge was specific: 'neglecting the business of his class by lecturing in a desultory manner and irregularly and by failing to supply subjects sufficient for the purpose of effective teaching.'[17]

Pattison was able to prove from his notebook that his lectures had been systematic and complete, and that he had procured so many bodies that he had even given some away. As an extra precaution, he then requested four of his medical colleagues to institute an independent enquiry. The resulting report, dated 9 December 1829 and received by the council on 12 December, was handwritten by John Conolly and signed by him and Anthony Todd Thomson, professor of materia medica and therapeutics, Edward Turner, and David D. Davis, professor of midwifery.[18]

During the enquiry, the four colleagues conducted several interviews among Pattison's students, concluding that 'the charge brought against our colleague is frivolous and destitute of foundation'. (There may, nevertheless, have been some substance in the students' complaint about the irregularity of Pattison's teaching. Apparently he was so given to the sport of fox hunting that he sometimes appeared in the lecture theatre with his gown carelessly thrown over his red coat and riding boots.) Further, they felt that the charges had originated in feelings that were 'too personal' in their nature to deserve the countenance of the council. 'We anxiously hope that in case of future complaints being made, the accuser will be called upon to support his accusations, and that we may be thus relieved from the degrading and insufferable conviction, that the character of each of us is at the mercy of every worthless, malicious and designing whisperer.'[19]

As 1830 began, Bennett was again complaining that Pattison had unnecessarily and unfairly made his lectures a prerequisite for admission to Bennett's demonstration course. Resenting this, some students had transferred to medical schools associated with the old London hospitals, which had been in competition with the medical faculty of the university from its very first day. At about the same time, Bennett demanded unsuccessfully that his title be changed from 'Demonstrator' to 'Professor of Practical Anatomy'; he ingenuously asserted that this 'would not interfere with any person's interest'.[20]

In February came an irritating little fuss about the university dispensary, the torchbearer for University College Hospital.[21] The council had considered that a clinical establishment under its direct control would be advantageous for the training of medical students. The dispensary opened on 28 September 1828, with John Hogg as the capable resident apothecary. Its object was to provide free medical and surgical assistance to the sick poor. Pattison was appointed as the dispensary's surgeon.

In a report of the special finance committee of the dispensary the accusation was made, based on the evidence of Hogg, that the medical personnel had been less than consistent in their attendance, and that Pattison was the worst offender. The council therefore recommended that a monthly record of the attendance of each physician and surgeon be kept, not by these individuals but by Hogg. A long formal protest by the three attendants (Conolly, Davis, and Pattison) was duly sent to the council in February. The authors deplored having their gratuitous services recorded and reported on by a subordinate, and the implied lack of confidence in their zeal and assiduity. Pattison's irregular attendance was acknowledged but justified by Conolly and Davis on the basis of the lack of surgical patients: 'When cases of importance have been offered to him, he has spared no attention, visiting the patients at their own houses and taking particular pains to interest the pupils in the progress of the disease or in the result of any operation which he has performed.'[22]

A few days later, Pattison sent two letters to the council on the same subject.[23] He pointed out that Hogg always knew of his whereabouts if an interesting case presented itself in his absence and that he spared no effort in treating serious conditions. ('When called on, I have cheerfully got out of bed, and gone to that distant part of the city to see a poor patient.') He added that a 'friendly hint' from the council would have prompted his regular attendance at the dispensary, 'even had there not been a single case to attend to'. Finally, he decried the way in which once again he had been accused on the basis of secret information and denied a rebuttal: 'God knows my whole heart has been wedded to the interests of the university, and to the best of my abilities I have endeavoured to promote them. Unfortunately, however, from some secret cause or other, my conduct has been the constant subject of suspicion and enquiry.'

From these comparatively minor scraps, Pattison's troubles rapidly assumed major proportions. The students' first public salvo was fired in a letter in The Lancet of 15 March 1830, signed by 'A friend to, and pupil of, the London University'.[24] The letter accused Pattison and Bennett of providing grossly inadequate coverage of their published course content and of generally slovenly and boring presentation. The language was at times abusive: Pattison's lectures were described as 'one long course of puff and nonsense' and a 'vile system of giving the lectures of anatomy'. One week later, 110 students signed a letter to *The Lancet*, completely vindicating Bennett but not Pattison.

The next attack was delivered openly by the ringleader of the dissident students, Nathaniel Eisdell,[25] who wrote two letters to the council complaining of Pattison's ignorance of and incompetence. in anatomy. He cited several instances of Pattison's errors and inadequacies and closed with the threat of public disclosure if his complaints were not settled. The council replied that their members did not think it prudent to institute an investigation on the representation of a single pupil. Eisdell promptly rounded up seventeen other students, who signed a statement supporting his original letter of complaint and demanding an investigation of Pattison's competence. They submitted their statement to the council on 15 May 1830.

At this point Pattison had no alternative but to appear before a committee of the council, 'although I felt exceedingly loath to come forward to defend my character from charges coming from pupils'.[26] Based on his testimony and that of three of his colleagues, Davis, Thomson, and Turner, a committee report, ordered on 12 June 1830, was duly submitted to the council a few days later. Pattison believed that he had disposed of the charges satisfactorily. The committee members thought otherwise. The report stated that Eisdell's complaints were partly valid, partly trifling, and partly incorrect; that the medical school was likely to suffer because of Pattison's increasing unpopularity; and that clashes of personality were exacerbating the problems in the medical school. The members felt that the truth of the situation would emerge only after a long investigation of a very difficult and delicate nature involving the testimonies of the students themselves. They recommended that the council instruct Horner to write to each of the seventeen students requesting statements outlining their grounds for complaint and that Pattison be requested to provide a written rebuttal of Eisdell's original charges.

The seventeen students were all out of town for the summer, so some delay was inevitable in receiving their submissions. But Pattison replied at once, in a letter to the council on 19 June. He began by enclosing a favourable testimonial signed a few months earlier by seventy-three of his students, in which they had acknow-

ledged Pattison's 'anxiety for their advancement' and 'the respect
they entertain for their Professor'. (They also expressed sympathy
for an attack of laryngitis which Pattison was suffering.) He noted
that among the names were ten of the seventeen complainants; he
found this change of view astonishing in such a short time. He
then systematically answered each of Eisdell's specific charges; for
example,

> *Complaint:* Upon a student asking him 'what nerves pass
> through the rectus externus oculi', he was unable to answer.
> *Answer:* I nor any anatomist living can be supposed to be
> prepared to answer on the moment the precise position of
> every hair-like filament of all the ramifications of the nervous
> system.
> *Complaint:* In demonstrating the peritoneum, he maintains the
> foramen of Winslow to be a hole in that sac.
> *Answer:* The fact is not so. But I can explain how an inattentive
> student may have been led into this error. [He then went into a
> discussion of the omenta and various views of the anatomists of
> the day, one of which, *if* correct, would have required the
> foramen of Winslow to be a hole in the peritoneum.]

Letters from all the seventeen students were eventually received
by the council and, to Pattison's surprise, they were sent on to him.
'And such letters! In one I am accused of not lecturing on a
particular organ; in another I am charged with lecturing too long on
the very same organ! One student says I was not minute enough in
the demonstration of the bones and another blames me for spending
too much of the session in the description of the skeleton.'[27] Most of
the students repeated Eisdell's original charges and many cited
Bennett's success as a teacher. Some students claimed that Pattison
started his lectures late and concluded them as much as fifteen
minutes early; that his style was monotonous and 'the impediment
in his speech made him perfectly unintelligible'; and that his lectures
were totally disorganized.

The council duly referred the students' letters to a special committee, the report of which, dated 2 July 1830, recommended that
some of the students be questioned in person by the committee,

with Pattison in attendance. Four of the seventeen students were selected and examined orally on 12 July.[28] Many of the complaints were repeated, particularly those relating to Pattison's poor attendance in the dissecting rooms and at the weekly revision sessions, and to perceived errors and omissions in his lectures.

Pattison was again required to reply in writing to these charges and this he did in a fourteen-page paper late in July 1830.[29] He pointed out how no other professor at the university had been subjected to any such investigation, whereas in his case 'God knows there have been enough of them'. Inconsistencies in the testimony were again shown to be blatant and frequent. These he enumerated at length; for example, some of the students claimed that certain topics had not been included, but then admitted that they had not attended the lectures at which these topics had been described. He asked the council to consider what the punishment of the complaining students would be if they had furnished such slanderous and conflicting testimony before 'any twelve men who could be packed in a jury box'.

Pattison was convinced that the students were being encouraged to lodge complaints by a person or persons in authority. 'A system of intrigue has been in operation against which no talents nor exertions could prevail.'[30] He considered none other than Leonard Horner, the warden, to be the guilty party and said so to the council in March 1830, making a formal printed statement in May. Details of the charges and countercharges were minutely detailed by both sides.* The dispute reached its climax the following year when Horner was subjected to such insults and innuendos by Pattison and Birkbeck that he withdrew his services and then formally resigned.

During the summer of 1830, Pattison was more and more frequently attacked in *The Lancet*. His most vociferous opponent was a militant demagogue named Alexander Thomson, the son of Professor Anthony Todd Thomson and a recent graduate of the University of Cambridge.[31] Under the pseudonym 'A Lover of Truth and Pupil of the London University', he outlined and supported Eisdell's charges against Pattison. Far more damagingly, he

* pp. 174–76 contain an account of this quarrel.

persuaded *The Lancet* to publish details of Pattison's feuds with Nathaniel Chapman in the August 1830 issue, which in turn led to further correspondence on the subject. The old troubles—the Ure divorce, the prostatic fascia problem, and Pattison's personal and professional difficulties in Philadelphia—were all raked up. In order to prove his unsullied reputation in the United States, Pattison arranged to have printed and circulated a large number of his testimonials.

Pattison was thus beset by students, colleagues, the warden, council members, and the press. In the hope of achieving some peace, he made a major concession to Bennett by arranging on 2 August 1830 for Bennett's promotion to 'Adjunct Professor of Anatomy'. Subject matter was to be divided between the two of them. It was decided that one or more prosectors were to be appointed to act as junior demonstrators and to prepare dissections for the two professors; that all certificates were to be signed by the two professors; and that students were to be allowed to attend either one or both of the sets of lectures. Pattison stipulated that announcements of the change should note that he personally had voluntarily suggested it. This was promised but not done.

Finding Bennett an excellent lecturer and demonstrator, the students championed him in the feuds with Pattison. Bell, who considered Pattison incompetent and damaging to the school, supported the students in their opinion of Bennett's superior ability. He was almost certainly more up to date in his knowledge and a more stimulating teacher than Pattison. He pressed for the teaching of general anatomy and recommended to his class a textbook by the French authors Bayle and Hollard. His introductory lecture in 1830 was the first formal acknowledgement in any English medical school of this new and popular branch of the science. After Bennett's promotion to adjunct professor, Pattison continued to teach descriptive while Bennett professed general anatomy.

Despite the praise heaped on Bennett by colleagues and students alike, Pattison from first to last remained at loggerheads with him. He claimed bitterly that Bennett might have had a hand in, and certainly approved of, Alexander Thomson's publication of Chapman's American account of the Ure divorce. He later accused

Bennett of encouraging the students to mutiny in 1831, and 'when he was lying on his death-bed, his hostility towards me ... was so inveterate, that he was in the constant habit of expressing himself in regard to me in terms of the most bitter malevolence and hate'.[32]

Bennett died on 27 April 1831 from tuberculosis of the bowel. Not all obituaries were kind: 'In the life of Mr. Bennett, the medical student may read a useful lesson on the passing folly of a restless ambition. He was ambitious and jealous of the fame of others. He partook of the sensitive constitution of the consumptive and was engaged in daily and nightly toils which such a constitution could ill bear.'[33] Pattison attempted to lecture on the day of Bennett's funeral, notwithstanding a general agreement among all the professors to declare it a day of mourning, but the students revolted and drove him, 'loaded with the execration of bleeding hearts, from his lecture-room'.

A few weeks after Bennett's death, the students, under the leadership of Eisdell, successfully petitioned the council for permission to sponsor a marble bust in his memory. They donated ten shillings each and by 15 July had raised sixty guineas towards the estimated ninety guineas required. The bust was duly completed and exhibited until its destruction by a bomb in the Second World War.

That Bennett was never answerable to Pattison was clear from the outset. Rivalry in seeking the students' affection and approbation was unavoidable. Bennett had sole control and direction of the dissecting rooms, from which Pattison was effectively banished. Pattison's perception that he was unwelcome there—'the field of my most useful operations'—may partially explain his poor attendance record, a record much denigrated in the various student memorials. Moreover, by giving regular, free unofficial lectures, Bennett was effectively underselling Pattison and promoting his own popularity. It was natural that the unfortunate Bennett, debilitated by a fatal disease, should accept the adulation of the students. And it was inevitable that Pattison, the students' idol in Baltimore and Glasgow, should resent his rival's popularity.

The Academic Year 1830/31

The academic year started with the surprising announcement that Bell, increasingly disillusioned with medical school affairs and the administration of the university, had resigned from the chair of surgery and that the vacated chair had been awarded to Pattison without applying for the post. Such a capricious appointment, made for no apparent reason, typifies the vagaries of the council at that time.

Before his resignation, Bell gave Pattison one final prod. During the summer of 1830, Bell had advertised widely that he would be giving 'lectures on the higher department of anatomy' in addition to those on physiology and surgery. Pattison had already been forced to give up to Bell the subjects of physiology and operative surgery and to Bennett half of the anatomy professorship; now Bell was claiming 'the higher department of anatomy'. 'I am fully aware of the foolish vanity by which Mr. Bell is actuated and since my connection with him, as a professor in the same university, I have for the sake of peace submitted to a thousand impertinences from him.'[34] Pattison then begged the council to publish an advertisement denouncing Bell's claim. The request was ignored.

Pattison was now reduced to being one of two professors of anatomy, but had acquired the chair of surgery. He was in more or less continual battle with the council and in particular with Horner, the warden. Dissension was growing among many of his professional colleagues, particularly Turner and Thomson. And the final, disastrous feuds with the students were rapidly gaining momentum. The Lancet alone devoted some seventy-nine pages to 'the Pattison affair' in the year 1830/31, and a four-inch pile of pamphlets, memorials, letters, and council minutes attests to the bitterness of the participants.

To make matters even more uncomfortable, the quarrels and inadequacies of the university were now being freely discussed in the newspapers. In the spring and summer of 1830, the Atlas, the Spectator, the Sun, The Times and the Morning Chronicle all published articles critical of the university's administration. On 5 July the Morning Chronicle gave a long account of a meeting of the proprietors, at which Birkbeck spoke at length in favour of an independent enquiry into the affairs of the university. He gave three reasons for its present deplorable reputation: the election of the disputatious Bell; Bennett's independent appointment in the anatomy department; and the appointment of Horner as warden. Birkbeck spoke strongly in support of Pattison's excellent qualifications: 'I believe Professor Pattison to be one of the best anatomical teachers in the metropolis—speaking in reference to his knowledge of the science—and perhaps the very best, speaking in reference to his perspicuous, animated and attractive mode of communicating knowledge.' He deplored the harassment Pattison had endured based on 'unsupported insinuations of ignorant students'. (*The Times* reported the same meeting in a more restrained manner.)

In addition to this humiliating public airing of his troubles, Pattison suffered a minor professional insult in connection with the anatomy textbook which he had recommended to his class, *The Anatomy of the Human Body* by his Scottish friend Andrew Fyfe (the last man in Edinburgh to wear a pigtail). In October 1830 *The Lancet* was highly critical of the book: 'miserable performance . . . can hardly speak in terms of sufficient reprobation . . . absurd blunders . . . the whole is so utterly bad . . . plates are merely sketches which would completely puzzle a student.' This was followed a month later by a bitter letter from a correspondent calling himself 'Montesquieu': 'A few weeks ago Professor Pattison introduced Fyfe's *Anatomy* to the notice of his class and bestowed some flattering lucubrations on the style and quality of the plates. . . . But for the timely review of this work in your able Journal, [many of Professor Pattison's] students might have been induced to purchase it upon the "*ipse dixit*" of an *imbecile professor of anatomy*.'[35]

But it was the students themselves who dominated the scene during the climax of the encounter. The ringleader in the early part of the academic year was again Alexander Thomson, a *force majeure* among the students, who had sent to the council during the summer a long and aggressive memorial, 'A Sop for Cerberus!', which included a fierce attack on Pattison. Its tone neither endeared Thomson to the members of council nor furthered the cause of the students. Pattison was charged with 'unusual ignorance of old notions and total ignorance of and disgusting indifference to new anatomical views and researches. . . . He is ignorant, or if not ignorant, indolent, careless and slovenly and above all, indifferent to the interests of the science. . . . Should you neglect our prayer, we warn you that we shall publish this very appeal. . . . Once more we urge you to dismiss this inefficient, careless, indifferent professor.' The memorial was later published.

Not surprisingly, Pattison and his supporters arranged for Thomson's expulsion. The stated reason was to prevent him from prejudicing the minds of the pupils of the new class. Not only was he banished from the university, but a letter was sent to Hogg at the dispensary, dismissing Thomson and forbidding him entry. Horner, who had been absent when the expulsion was ordered, rescinded it after Thomson had promised that he would not interfere with Professor Pattison. Yet, on the very day that he was re-admitted, Thomson went to Pattison's lecture theatre and mounted another bitter attack before the students. For this he was again expelled— and again re-admitted two days later, after writing an abusive letter to Horner in which he referred to the college servants as 'hireling slaves'. With Horner's permission, he addressed Pattison's students at noon on 13 October after Bennett's lecture, with Eisdell acting as chairman. The proceedings lasted two hours and Eisdell vacated the chair only when Pattison arrived to give his lecture at 2.00 p.m. Pattison informed the council, and for the third time Thomson was expelled. This time he received no redress and shortly afterwards he moved to Paris, from where, about a year later, he wrote a long, impassioned tract to Lord Brougham, chancellor of the university, about the whole matter.

The damage had been done. Scenes of convulsion and violence ensued; parties and cabals were formed; and tri-coloured papers inscribed 'Thomson and Liberty' were handed about. At this point any semblance of discipline vanished from Pattison's class and his life became intolerable. The students blamed him for seeking to silence criticism by the expulsion of his critic. Attendance dropped and matters became steadily worse. By February 1831 the students were flagrantly disobeying orders and using insulting language to the college servants. In the same month they submitted yet another memorial to the council, couched in beautiful if obsequious language and signed by sixty students. While acknowledging 'the once kindly bearing of Professor Pattison . . . and his urbanity of manner', the students repeated their earlier accusations of unsystematic lectures, uninteresting presentation, superficial subject matter, and frequent errors. They requested yet another enquiry.

At the same time Pattison announced new rules of discipline. Henceforth students would be required to sit in the body of the theatre and not up on the back bench usually occupied by the rowdiest members. A daily roll call was to be introduced to ensure regular attendance. (Ironically, only two years previously Pattison had condemned the practice of calling a roll: 'Medical students are a very difficult class to manage and to attempt to treat them as schoolboys would have the effect of driving the diligent, as well as the able, from our establishment.' He had considered it the professors' duty to make their lectures 'so interesting as to induce the pupils to consider it a loss to themselves to miss a single lecture'.[36]) In addition, every student who wished to obtain a certificate would be required to submit to regular weekly examinations. Pattison freely admitted that he hoped, at the weekly examination, to expose the ignorance of the ringleaders of the dissident students, in order to prove how ill qualified they were to pass judgment on him.

Enforcing such regulations proved difficult. On the day following the announcement, six students occupied the prohibited back bench and contemptuously refused to move down. Pattison appealed to the council, which approved the new regulations and recommended that beadles (university law enforcement officers with constabulary powers) be summoned in future instances of disobedience. The following day thirty students again occupied the back bench. Pattison reported the council's recommendation that, if necessary, 'they be dragged down' by the beadles; twenty-seven students still remained seated. Fearing a riot if the beadles were called in, he merely recorded the students' names and reported them to council. Suspension of the twenty-seven followed and council met to decide further punishment. Vacillating and inconsistent again, the council readmitted the students provided they apologize—not to Pattison, but to Horner. The students readily complied.

Further insubordination soon followed. Arriving for his next lecture, Pattison found a student, Merriman, sitting on the prohibited bench. He asked him two or three times to come down, whereupon Merriman got up and asked 'in the most insolent manner possible, if Pattison asked him to do so as a favour'. Pattison replied, 'If you put it on that, I require you, as your professor, immediately to come down.' 'Well I won't.' Bree, another student, rose from his place in the theatre and scrambled over the benches to join Merriman. The class cheered.[37]

It so happened that a proprietor of the university, Herbert Fearon, had been requested to attend one of Pattison's lectures as an observer; the request was made by an anonymous medical student, who referred to Pattison's lectures as 'a serious evil'. Fearon duly attended the very lecture at which Merriman and Bree were misbehaving. His report to the council provided an independent description of the anatomy lectures:

> There were present about 120, the great majority of whom appeared to be attentive and studious. A few, who occupied the higher seats, were in their conduct the reverse of either, being apparently wholly intent upon other subjects. . . . In the calling over the names of those present, some half dozen students left their seats and placing themselves near the door, answered in a tone of derision to nearly every name which was called over. The announcement of a lecture on surgery for four o'clock on the same day was received with general applause, with the exception of the party near the entrance, who were loud and pertinacious in hissing.[38]

By coincidence, the council received a day or two later, on 5 March 1831, a memorial signed by forty-one students, testifying to Pattison's 'punctuality of attendance, his assiduity and ability as a teacher and his conduct as a gentleman'. (One of the signatories, W. R. Jones, later turned against Pattison in Philadelphia.) A later memorial of complaint carried eighty-nine signatures, showing that the class was divided about two to one against Pattison.

Early in March 1831 Pattison wrote three more letters to the council, describing further acts of insubordination. Bree had changed his tactics on the day following his misbehaviour with Merriman. On this second occasion, he sat in the front centre seat of the theatre and read a newspaper with great deliberation during the whole of the lecture. When the lecture had ended and Pattison had returned to his private room, another of the actively insubordinate students, Peart, approached him demanding a certificate of attendance. Pattison who knew of his extremely poor attendance record, refused the request. Peart's friend Edward Meryon (the gold medalist in Pattison's anatomy class of 1829/30) stepped forward and said, 'Mr. Peart has attended you as regularly as anybody else.' Pattison replied that he had no desire to have any more conversation with them. Peart then said, 'You are no gentleman. Your conduct is most ungentlemanly and unmanly.' He retired to the classroom where he recounted the episode to his friends. On being asked by one of them how Pattison had reacted, 'Oh,' he said, 'Pattison took it very quietly.'

Peart immediately complained to the council. Pattison's reply to the council's enquiry centred upon Peart's poor attendance record and general ignorance of anatomy, but he offered to provide the certificate if Peart submitted to and passed an examination. Peart thereupon wrote a long letter to the council, pointing out that he did not seek a certificate of competency but merely one of attendance. He complained that Pattison's real motive was a vindictive desire to make an example out of one of his most intractable opponents and forthwith instructed his attorney to serve a writ on Pattison to force him to grant the certificate. This was reported to the council by Pattison's solicitor, with a request for instructions on how to proceed.

Following Pattison's complaint about the behaviour of Merriman, Bree, and Peart, all three were suspended, but, on 7 March, Bree attempted to gain entrance to Pattison's lecture by force. He struggled with the doorkeeper, who had been ordered by the council to keep him out, remarking, 'I do not give a damn for the lectures, but, by God, the council will not keep me out.'[39] He was encouraged in this behaviour by the shouts and laughter of the other students. Having tried unsuccessfully to lecture, Pattison retired after twenty minutes, vowing not to return until after the council had met. Pattison's patience was wearing thin. He stated, 'My lectures were formerly my greatest pleasure, but I now enter my classroom with a feeling of loathing, feeling that I am to be insulted, and that no zeal nor exertion on my part will obtain for me that approbation to which I know I am entitled.'[40]

The council reprimanded Merriman but allowed him to continue attendance at all lectures in the medical school. Bree and Peart were expelled from Pattison's lectures for the balance of the year, but after they had apologized the order of expulsion was rescinded. Peart then withdrew his action, whereupon the council ordered the certificate of attendance to be granted. It was obvious that the council had administered the *coup de grâce* to such little authority as Pattison still possessed. A committee of the students now met daily on the university grounds and decided whether or not he should be allowed to lecture. This 'standing committee of the students' continued for the remainder of the academic year and was even recognized by the council, with whom they corresponded. While these events were unfolding, they were given full publicity by debates filling the columns of *The Lancet*.

A state of open rebellion and riot now existed in the anatomy classroom. The well behaved students sat in the lower area, while the rebels sat in the upper part, including the prohibited back bench. The theatre was often filled to capacity by students from other faculties who had come 'to see the fun'. Shouts, howls, hisses, yells, and whistles abounded. Cries of 'Off, off! You won't be permitted to lecture' were heard. Pattison daily tried to make himself heard but was always forced to retire.

Daily reports were sent to the council, which finally appointed a commission to visit the theatre on Wednesday, 16 March 1831; the events were described in the *London Medical Gazette* of 26 March. The usual uproar was in evidence when Horner entered. The noise and clamour then became deafening and he was not allowed to speak. The door opened again and three members of the council, headed by Lord King, presented themselves. Lord King, after humbly begging the students' attention, succeeded only in promising that all complaints submitted to council would be reviewed immediately. At this, a student rose and declared that the class had already sent in a memorial signed by sixty students which had been ignored. Clamour again prevailed and his lordship, with his party, was obliged to retire, 'leaving the pupils triumphant and affording an excellent illustration of collegiate discipline'. The students then submitted a verbatim copy of the memorial, now signed by eighty-nine students, thereby taking Lord King at his word.

Full power was now vested in the standing committee of the students. Pattison stated, 'I never knew, in fact, when I went to the university, whether I should or should not be permitted to lecture, until I entered my classroom and saw the disposition of the students.'[41] This state of affairs continued for a month until the end of the academic year.

In an attempt to obtain independent support for his competence as a lecturer, Pattison proposed to the council on 23 March that a reporter be present at all his subsequent lectures, so that 'a faithful report of what I *really* do teach be obtained'. He offered to pay one half of the expenses involved, but stipulated that he be furnished with each report for the correction of any errors prior to its submission to the council. The proposal was rejected. Another measure ·to secure order was passed by the council in April. This required all students to sign a paper promising good behaviour, while those who refused were to be excluded from Pattison's lectures. This sensible move came too late to have much effect.

In the meantime, Lord King had been fulfilling his promise to attend to the memorial of the eighty-nine students. He requested them to submit in writing specific charges against Pattison. John Rayner, the secretary of the standing committee of the students, initially declined to comply, suggesting that a *viva voce* examination would be more appropriate. On being pressed for this information, the students provided, on 16 April, a detailed eleven-page report, listing charges regarding anatomy and surgery; a subsidiary list of sixty-five topics allegedly omitted from the surgery lectures was sent to the council by three other students.[42] The charges in the first of these were specific: for example,

1. An evident want of a liberal education, frequently evinced by the commission of flagrant classical errors.

   Examples. Ductus communis chole*doctus* [for choledochus], corpus spongiosum urethr*a* [for urethrae], pub*ises* [for pubes], anastomos*ises* [for anastomoses], extravasat*ated* [for extravasated].

2. An inadequate acquaintance with the scientific and practical departments of those subjects which he professes to teach.

   Example. When giving the anatomy of the femur, he went over the attachments to the great trochanter three times and each time in a different manner: first, that the gluteus maximus is inserted into it; second, that the gluteus maximus and medius are inserted into it; and third, that it receives the insertions of the gluteus medius and minimus. [The last statement is correct.]

3. A want of methodical arrangement in lecturing on any particular subject, with great neglect of the relative connections of parts.

   Example. In demonstrating a muscle, its origin and insertion are frequently indicated by the phrase 'the part which I now touch' instead of specifying in words the exact names and situations of those parts. The points touched can only be seen by a few and, besides this, it is unscientific and indicates a deficiency of powers of expression.

4. The superficial and careless manner of treating many, and total omission of the consideration of other important subjects.

   Examples. The important subject of gunshot wounds occupied but three quarters of an hour. The diseases, injuries and accidents of the spine had a quarter of an hour allocated to them.

5. The non-redemption of pledges given to his class.

   [As an example the students cited a promise made by Pattison in his introductory lecture of the 1830/31 session that he and Bennett would provide two complete courses of anatomical lectures.] This pledge has not been fulfilled by Professor Pattison.

Armed with this information, a committee of the council met to examine the question of Pattison's competence. Pattison himself had little faith in this committee:

> The injustice of the acts of this committee is only exceeded by their absurdity. Without taking the pains to ascertain whether I had really committed the errors with which I was charged, they proceeded to investigate whether the charges made did or did not contain anatomical blunders. Lord King [and others], not one of whom knew a nerve from an artery, constituted themselves the judges of my anatomical pretensions!. . . With [anatomical engravings] before them and with the assistance of anatomical dictionaries to explain technical terms, these gentlemen gravely deliberated on the amount and correctness of the anatomical knowledge possessed by the professor of anatomy.[43]

At the end of three weeks the committee submitted an inconclusive but damaging report, as a result of which, at the council meeting of 30 April 1831, one of the members gave notice of motion 'that it be recommended to Professor Pattison to retire from the professorships of anatomy and surgery at the close of the present session'. At 5.30 p.m. on the same day, Horner sent him a note informing him of this development. Pattison's career at the University of London was now effectively at an end.

On learning from Horner about the notice of motion, Pattison sent a five-page letter to the council, pleading that his dismissal would lead him 'to beggary and ruin':

> I have been told by a member of council that, although nothing has been established against my competency, still my reputation has suffered so much from the investigations that if I remain in the institution, the medical school is greatly perilled. I admit that this is true and I feel that if I remain, it will require many years for the school to recover from the injury it has sustained. But surely the fault did not rest with me. I did not, as you are aware, originate these proceedings, which have led to a result so unfortunate, nor did I, by my conduct, furnish any just cause for others to do so.[44]

Much activity ensued on the part of the council, the proprietors, and Pattison's colleagues. Conolly had already resigned, protesting that the action of the council not only was unjust to Pattison, but also posed a threat to each of his colleagues. Six professors from other faculties sent a memorial to the council with the threat of resignation: 'We beseech the council to bear in mind that injustice can never be defended by expediency; that whatever be the effect of the proposed measure, it must in the long run be followed by evil, which never fails to attend a departure from strict equity.'[45]

After several delays and much painful deliberation, the council brought the conflict to a close on 23 July 1831 by adopting a resolution, smacking of sophistry, that called for Pattison's dismissal but with an untarnished reputation:

> Pursuant to his notice of the 9th inst., it was moved by Mr. Bingham Baring, seconded by Mr. Henry Warburton:
>
> That the council in concurrence with the suggestion contained in the report of the select committee of the council of the 18th June, 1831, 'that the popularity and efficiency of the medical school had received a shock by the disturbances which have prevailed in it and which can only be obviated by the retirement of Professor Pattison from the chairs of anatomy and surgery'; and deeming it therefore essential to the wellbeing of the university and the success of the medical school that Professor Pattison should not any longer continue to occupy those chairs; resolve:
>
> That Professor Pattison be, and he is hereby removed from his situations of professor of anatomy and surgery in the university.
>
> That, in taking this step, the council feel it due to Professor Pattison to state, that *nothing which has come to their knowledge with respect to his conduct has in any way tended to impeach either his general character or professional skill and knowledge* [my italics].
>
> Which motion having been put, there appeared in favour of it thirteen, two members declining to vote; and it was carried accordingly.

Leonard Horner, warden of the University of London, engraved by
Thomas Dick after a portrait by T. C. Wageman. Photo by Tom Scott.
(Scottish National Portrait Gallery)

Let the clerk transmit to Mr Pattison a copy of the resolution.
Adjourned.

Zachary Macauley (signed) confirmed: Thomas Coates[46]

Pattison's dismissal was now final and the only question remaining was how much compensation, if any, he was to be awarded. In June he wrote to the council stating that any mutually agreeable settlement had to be accompanied by a formal statement that no charge had been established against him. Birkbeck suggested £400 per annum, Davis £300, and Grant, Turner and Thomson 'a reasonable sum'. After several months, Pattison wrote to the council, on 23 November 1831, requesting a speedy decision, as he had been elected to a professorship in the United States. By his abrupt dismissal he had been deprived of one year's income and was still owed the unpaid salary which he had earned in 1827/28. The council eventually approved a single lump sum of £200, which Pattison accepted 'as a full discharge of all my claims upon the university and as terminating all discussion and differences with it'.[47]

Leonard Horner, the Warden

Throughout Pattison's difficult years at the University of London, one of his chief sources of irritation was the warden, Leonard Horner. Formerly a linen manufacturer in Edinburgh, Horner took an exalted view of his office. After considerable quibbling about his salary and title, he finally settled for £1200 and the title of 'warden' (having rejected the original proposal of 'secretary').

He was undoubtedly well intentioned and of a kindly disposition,[48] but he appears to have been weak, biased, and vacillating and was frequently guilty of accepting and acting on ex parte testimony. Although remote in his personal relationships with the professors, he was well regarded by many, being punctilious and a strict observer of the proprieties. Lord Cockburn (the Scottish judge) had thought him 'one of the most useful citizens that Edinburgh ever possessed'.[49]

His very generous salary became a not illegitimate grievance to the poorly paid professors, particularly when he was known to be unlearned. They were aware that it was paid from general management funds which in turn were supplied by each professor, who contributed one third of his meagre income. Indeed, Horner's salary was seriously questioned in public: the *Spectator*, for instance, remarked on 13 February 1830: 'The thing is absolutely ludicrous; if he gets £400 a year he is well paid.'

Communication between the council, the professors, and the students was the responsibility of the warden, and he alone was in a position to influence the decisions and actions of the council. He soon proved to be guilty of petty interferences in the affairs of the professors and of perpetually referring every trivial dispute to the council. In 1830 he committed a serious indiscretion in writing an anonymous letter to the *Sun*, his identity being exposed five days later. In this letter he admonished the professors to 'labour earnestly to establish their reputations' and added that 'the sooner that is done, the more rapidly will their emoluments be increased'.[50] Such arrogance did not sit well with professors who were receiving salaries of £150 to £400 and who were already recognized as eminent scholars in their fields; Horner was later rebuked by the council for his action.

Within the walls of the university, a group of nine professors, including Pattison, pressed in May 1830 for the abolition of the office of warden, on the grounds that Horner was merely the clerk of the council, that the professors should transact directly with the council without the intervention of any intermediary, that the salary granted to the warden was an unjustifiable drain on the professors' income, and that the release of his salary would allow a general reduction in student fees.

The professors' proposals marked the start of a concerted attack on the warden, which at times became confused with Pattison's defence. As matters became steadily more bitter in 1831, so also did the accusations against Horner. Birkbeck was Horner's most vitriolic and outspoken opponent in the body of the council and Pattison's most powerful ally. During the previous year, Pattison and Birkbeck had accused Horner of receiving Eisdell in the council room, a charge which Horner denied but which Eisdell subsequently confirmed. Horner was therefore vulnerable to the imputation of lying and of encouraging Eisdell of inciting his fellow students to rebel. On 26 February 1831, Birkbeck provoked Horner again by suggesting that this time he had encouraged Peart in his actions against Pattison and that his receiving money from the university 'deprived him of the privileges of a gentleman'.[51]

Soon afterwards, Pattison, in a long letter to the council, openly accused Horner of complicity in the revolt of the students. After a fifteen-page preamble, he made four formal charges: that Horner had re-admitted Alexander Thomson after he had been expelled and that he had repeated the offence after Thomson had been expelled a second time; that Horner had actually drafted a memorial for the students to copy and send to the council, with the objective of initiating an enquiry into Pattison's competence; that Horner had told a student that Pattison was exceedingly foolish in requiring the students to sit in any particular seat and that 'the thing would soon be put a stop to'; and that Horner had frequent conferences with Peart in the council room after the latter had called Pattison 'no gentleman', and that he encouraged Peart to persevere in his misconduct. All the charges were supported by the evidence of independent witnesses.[52]

On 23 March, Horner duly provided the council with a rebuttal of the four charges. The first charge he did not deny, but provided lengthy arguments to prove that Thomson had convinced him of his innocent intentions on both the occasions of his re-admission. The second charge was also partially acknowledged, but Horner's intention had been to keep the dispute within the walls of the university. He denied the third charge and obtained a letter from the student in question who called Pattison's charge 'a gross invention'. Regarding the accusation concerning Peart, Horner admitted having interviews with him, but claimed they concerned only Pattison's refusal to grant Peart a certificate.

The accumulation of charges and innuendos, coupled with a total failure of the council to provide him with support, caused Horner to absent himself from the council meetings for two weeks and then to submit his resignation on 26 March 1831, 'fairly scared and worn out by vexation and injustice'.[53] When pondering on the words he should use in announcing his resignation to council, Horner asked

Bell for his advice. 'My dear Leonard,' was the reply, 'stand up, show yourself, and say, "Gentlemen, I came to your university comfortable and well filled up—look at me now, shrivelled and thin, my clothes a world too wide." That would be true eloquence.'

Horner seems to have lacked the ability to look below the surface in assessing colourful if unorthodox colleagues. He was ever ready to investigate a new accusation against Pattison but unwilling to follow up expressions of support. He was frequently influenced by his friends, notably Bell, who supported Bennett to Pattison's detriment. And he was continually harried by his opponents, notably Birkbeck, who supported Pattison out of personal enmity of Bell. Further complications arose from the ill feeling and jealousy of the professors from other faculties, who therefore supported Pattison as a means of attacking Horner.

Pattison blamed Horner for most of his troubles—for Bennett's independent appointment, for scheming with Bell in 1829 to remove Pattison from his chair, for failure to make due reparation for all the early investigations into Pattison's conduct, for his support of Eisdell, Peart, and Thomson, and for manipulating the council to further his own ends. Horner in turn considered many of the petty complaints which flooded his desk as trivial and irritating. For this reason, he decided in a completely arbitrary manner which of the complaints merited the attention of the council. He took upon himself the responsibility of setting the beadles to take student attendance records, and he fussed and interfered with the purchase of equipment and apparatus. All these, and the assumption of the position of literary head of the university, were done without the warrant of any scholastic prominence.

It is ironic that the three main actors in this violent Pyrrhic drama should all leave the stage in 1831: Bennett to his Maker, Horner to Bonn with a nervous breakdown,[54] and Pattison to a humiliating temporary obscurity.

The Aftermath

Pattison's dismissal caused immediate repercussions. Three professors from other faculties resigned, including Augustus de Morgan, professor of mathematics, who made his views plain to the council: 'Here is distinctly laid down the principle that a professor may be removed and, as far as you can do it, disgraced, without any fault of his own. This being understood, I should think it discreditable to hold a professorship under you one moment longer.'[55]

It must not be supposed that such a *cause célèbre* would fade away without a great deal of discussion. The newspapers followed developments avidly, including Pattison's two last unsuccessful appeals to the proprietors in August and September. The *Morning Chronicle* of 22 August reported the extraordinarily naïve statement of Lord Ebrington, the chairman of the council: 'In the university there was no power over the students, and therefore the professors must be popular or the university could not flourish. It had been proved to the council that Professor Pattison was not popular, and therefore the council thought it was not discharging its duty to the institution and public, in not dismissing Professor Pattison.' This prompted the paper's observation: 'The pupils were in a state of mutiny, and the council, who were in a state of imbecility, instead of punishing the students, punished Professor Pattison.'

In August 1831 Pattison published an account of the affair, *Professor Pattison's Statement of the Facts of his Connexion with the University of London,* from which some of the above narrative has been reconstructed. Earlier, he had informed the council of his intention to do this, and requested from them copies of all the relevant minutes, letters, memorials, and other documents, to ensure maximum accuracy; the request was refused without explanation. Before circulating the statement, he sent it to Birkbeck, who deemed the work 'substantially and generally minutely correct and . . . remarkably free from exaggeration'.

While the *Morning Chronicle* in particular had championed Pattison's cause, Thomas Wakley, the founder and editor of *The Lancet,* took the opposing view. An avowed opponent of Pattison, he mounted a bitter attack in a series of unsigned leading articles in 1831. A lifelong agitator and zealous reformer, he was against any seeming monopoly of men or institutions. He criticized Pattison's lecturing style: 'He appears to have no idea of modulating his voice—not at best an agreeable one. . . . He begins his lectures in a low tone, rises rapidly to the top of his voice, and delivers the remainder as loudly as he can—we were about to say scream. . . . This may have been a reason why at least *some* of the students preferred to sit on the most remote benches.'[56] Then he referred to Pattison as 'a bad anatomist and a still worse man', and gave a biased account of his life and attainments, alleging that Allan Burns's association with him resulted merely from Burns's fondness for one of Pattison's sisters.

Pattison used the pages of *The Lancet* to answer Wakley's attacks. The accusation concerning his sister Margaret's association with Allan Burns he demolished in no uncertain terms: 'I shall not condescend to reproach you for the indelicacy of thus gratuitously dragging a lady's name before the public. . . . Mr. Burns was a gentleman of the most retired habits and was only once, to my recollection, in my mother's house. . . . I may further observe that that young lady was, at the time to which you allude, a mere child.'[57]

Pattison's *Statement* provoked further public protestations by supporters and detractors alike in the pages of *The Lancet*, including letters from Eisdell and from one of Bennett's sisters. Turner and Thomson addressed a pamphlet to the proprietors, giving their side of the dispute. It explained the reasons for the change of their original friendship and support of Pattison to their later enmity and criticism. As well as reviewing all the past events, Turner questioned Pattison's ability: 'Mr. Pattison has been a teacher of anatomy and physiology for I believe about twenty years, and yet I greatly doubt if Mr. Pattison can lay his finger on a single fact and claim it as his contribution . . . to the stock of anatomical and physiological science. It we enquire whether he is the author of any useful compendium for facilitating the acquisition of anatomy, we shall find an equal blank.'[58] Pattison could have responded by alluding to his previous outstanding reputation as a lecturer, to his achievements at the Baltimore Infirmary, and to his vigorous and successful efforts in support of the Anatomy Act. Instead, he replied with another lengthy account of the events leading up to his dismissal.

The last major commentary came from Alexander Thomson in the form of his 144-page pamphlet addressed to Lord Brougham, chancellor of the university. It is a fascinating tirade, dripping with venom and injured innocence, written in the aggressive and exaggerated style of an aggrieved young man of those days. On the title page he referred to Pattison as 'Mr. Pattison, a discarded if not disgraced Professor', and throughout he did not mince words in defence of the students' cause. Referring to a condescending remark in Pattison's *Statement*, he enquired: 'Pray, what more is a professor than the temporary, hired servant of the pupil?' He ended by ridiculing Pattison's alleged pride in the Cadwalader duel, since no challenge had been issued during the London disputes: 'Why sought he neither legal nor *honourable* redress? The pistols are always ready; they fearfully decorate his drawing-room table; but I fear they have ceased to be much used; . . . the HERO now contents himself with paper bullets.'

Factors in Pattison's Dismissal

The students were a powerful force in the university: their attitudes and opinions governed the popularity and prestige both of the professors and of the institution itself. The number of students attracted to a classroom was influenced as much by the personality as by the learning of the professors. And the greater the number of students, the greater were the rewards for all concerned. This then was the prime reason for the council's anxiety over Pattison's unpopularity in London: news of an empty lecture theatre was a bad portent for the financial stability of the university.

There was a certain air of militancy and rebellion among the students, who brooked no pomposity or incompetence. They were intolerant of Pattison's course content and teaching methods. The anatomy he taught was dry, classical, and descriptive and differed little from courses given fifty or sixty years earlier. The students were aware that exciting new discoveries were being made which they felt Pattison ignored. His recent contempt for the new and popular general, or French, anatomy was a further cause of criticism. ('I am complained of, I am told, because I do not teach "French anatomy". This is a new phrase; and I would ask, in the name of common sense, what is meant by it? . . . I teach anatomy for the purpose of educating useful medical practitioners.'[59]) Bennett's espousal of the new approach and Wakley's support for Bennett magnified the discord.

Another factor which worked against Pattison was his nationality. The strongest influence in the university was Scottish (Bell, Birkbeck, Conolly, Horner, Pattison, Turner, and Thomson were all Scots), yet Scotsmen were not popular in London in those days. Pattison was taken to task by the students for recommending an anatomy textbook by a Scottish author; Bennett apparently entertained the students by mimicking Pattison's Scottish burr; and Wakley in *The Lancet* made fun of Pattison's nationality: in referring to his most loyal supporter on the council, Wakley imagined Pattison saying, 'I will away to my friend Birkbeck, who, being fra' the nooth, will stick closely to a brither Scot.'[60]

The council must also bear its share of the blame. It was too large to be effective, and its members were sometimes not interested in university affairs and always lax in attendance. This left Horner in the position of having to make rapid decisions without the formal authority to act. At times the council seemed more ready to hear and act on submissions by the students than on those from the professors. The professors' lack of voice in the council seriously impeded their co-operation. 'A professor in [this] institution is on the same footing . . . as a domestic servant to his master, with however the disadvantage of the former not being able to demand a month's wages or a month's warning.'[61]

It appears, then, that Pattison's troubles at the University of London were the result of the interaction of several individuals and groups: Pattison himself, conceited, antagonistic, 'old school', and a Scotsman; Bennett, the very capable, independent demonstrator of anatomy; Horner, the weak, unscholarly warden; the large, unreliable, inconsistent council, the members of which conferred no authority or administrative responsibility on the professors; proprietors who failed to support or regulate the council; students who were ready and willing to censure incompetence or injustice, particularly when supported by senior members of the university; professors who were consumed by animosity and jealousy; and, lastly, Wakley, the editor of *The Lancet*, who was opinionated and powerful.

The quarrels had several lasting consequences. The chair of anatomy was eliminated. Pattison's approach to teaching fell into disfavour, and his course content was divided between the departments of what afterwards came to be called physiology and descriptive anatomy. The constitution and administration of the university were likewise changed. The office of warden was abolished and replaced by a secretary at a salary of £200 per annum. The council appointed a committee of seven of its members to conduct the ordinary business of the university, with the whole body meeting at less frequent intervals. The regulations governing the tenure and powers of the professors were regularized. A *Senatus Academicus* was established; its president was a council member, ensuring improved liaison between the academic and administrative branches of the university.

Pattison's three years of vexatious labour undoubtedly contributed to these improvements, but his troubles were recalled unsympathetically for many years to come. Ten years later, a physician in Bristol wrote to Pattison's colleagues in the United States: 'I wish his colleagues joy of Pattison. He did all but ruin our so-called London University . . . during his brief connection with it.'[62]

Departure from London

About two months after his dismissal, Pattison was appointed to a professorship at Jefferson Medical College in Philadelphia. He had the satisfaction of announcing this to the university council in the course of other correspondence. News of the violent upheaval at the university had already crossed the Atlantic; a letter written in Philadelphia in October 1831 noted: 'We hear in this country a very poor account of the London University. Professor Pattison has been invited to Philadelphia, where I am sure he will find a snug two thousand sterling a year.'[63]

He had wisely maintained a good relationship with his friends and colleagues in the United States and was always willing to act on their behalf when so requested; in 1828, for instance, he provided a report on the candidacy of the surgeon George Bushe for a faculty position at the Rutgers Medical College, New York. Moreover, he had been a consistent supporter of American medical achievements. Responding to Turner's comment that his earlier successes in the United States did not guarantee success in London, Pattison had replied:

> The Americans, forsooth, are very easily satisfied. It is very well for John Bull to say so. I should, however, have expected that you were too well informed as to the state of medical science in the United States, to have given currency to such a vulgar and unfounded prejudice. The Americans, I assert, are as far advanced and as enlightened in their medical opinions as any country in Europe. . . . Let the improvements in medical and chirurgical science be reviewed for the last thirty years and it will be found that America has furnished her full quota.[64]

Pattison spent Christmas 1831 with his mother in Glasgow and remained with her there until April 1832 because of her failing health. While in the north, he no doubt read in *The Lancet* that about 120 students of the medical school of the University of London had entertained their professors on Friday, 23 February 1832. 'Mr Nathaniel Eisdell, a distinguished student of the institution, acted as president. . . . The memory of the late Mr. Bennett was respectfully and appropriately proposed . . . and drunk in solemn silence.'[65] Wakley commented that the occasion was the first dinner of the kind ever given by medical students in London, and that it spoke well for the good feeling that prevailed between teachers and pupils at the university.

Having settled his affairs, Pattison set sail on board the *Napoleon* on 24 April 1832, bound once more for the New World.

VI
Relative Tranquillity
in the United States
1832–1851

BY THE TIME Pattison set sail for the United States to take up his appointment at Jefferson Medical College in Philadelphia, most of the major crises of his life were over. He had lived forty-one far from tranquil years; as might be expected, the remaining nineteen would also bring feuds, arguments and dissension, but none on the scale of those past.

The picture that emerges from previous chapters is of an aggressive man, ever ready to indulge in physical or verbal combat. His opponents considered him to be arrogant, vain, conceited, and antagonistic, while recognizing his ability to rebound, apparently unmarked, after each adversity. His supporters, on the other hand, extolled his brilliance as a teacher, his kindliness, and his generosity.

Such divergent views would later be echoed by Pattison's colleagues at Jefferson Medical College, where he arrived in July 1832 as professor of anatomy. But, for the time being, he had left controversy and the humiliation of his dismissal from the University of London behind. Quickly he immersed himself in his new responsibilities.

Jefferson Medical College

The Jefferson Medical College, Philadelphia (now known as the

183

Thomas Jefferson University) was founded in 1826 in opposition to the powerful and prestigious medical school of the University of Pennsylvania. Its founder, George McClellan, a skilled surgeon and teacher, had built up a successful private medical school which he sought to incorporate.

The university's monopoly was threatened and its officers, exercising their political strength, succeeded in blocking legislative approval. McClellan then chose to ignore the university and the vested interests of its professors by arranging that his school be a college under the aegis of the Jefferson College at Canonsburg, Pennsylvania—a Presbyterian college of the arts and sciences, established in 1802. The latter's trustees duly obtained a charter from the legislature to grant medical degrees, thereby putting the medical college on an equal footing with the university. (In 1838, the college received an independent charter under the name 'The Jefferson Medical College of Philadelphia'.) It is not surprising that from the outset there was little love lost between the two competing institutions.

In the *Annual Announcement of Lectures* for 1832, Pattison praised the college with characteristic bombast, comparing it favourably with the oldest chartered medical schools. He pointed out, for example, that its newly constructed dissecting rooms were unsurpassed by any in the world; he drew attention to the large number of students who had transferred from the University of Pennsylvania; and, to encourage the trend no doubt, he urged prospective students to attend the free introductory lectures given at both institutions before deciding at which to enrol. Such *Announcements*, an essential feature of all American universities and colleges at this time, not only provided factual material about courses, fees, and facilities, but also served as a platform for extolling the advantages of the institution. In his contributions to many of these, Pattison was often guilty of exaggeration and pomposity.

Students paid fifteen dollars for each of the six courses of lectures given by Samuel Colhoun (materia medica), Jacob Green (chemistry), George McClellan (surgery), his brother Samuel McClellan (midwifery, which included diseases of women and children), Pattison (anatomy), and John Revere (theory and practice of physick). A further fee of five dollars was to be donated to the janitor, who undertook to provide for each graduate 'a handsome box for the preservation of his diploma'.

Pattison had much to say in the *Announcement* about the importance of dissection:

> Anatomy, the basis of all medical reasoning, can only be studied *practically*. . . . Should [the student] neglect his opportunities for acquiring a competent knowledge at College, he must be content to continue forever afterwards a mere driveller in his profession. Now, anatomy is not to be learned by an attendance on lectures. Dissection, and dissection alone, can make a man an anatomist. The professor of anatomy, it is true, may, by animated and masterly demonstrations, do much to assist the anatomical student in the prosecution of his studies, but it is in the dissecting room, with the dead body before him, by patient and assiduous dissection, that the student can alone acquire a knowledge of anatomy.

Recognizing that the students needed a good comprehensive anatomical museum to complement the knowledge attained by dissection, he proposed that each should prepare labelled anatomical and pathological specimens as the start of a permanent collection. As an inducement to the students he suggested that the specimens be considered as memorials of their connection with their Alma Mater and as memorabilia for their sons to emulate or possibly surpass. The plan was successful and a curator of the college museum was appointed in 1834. Finally, to bridge the gap between dead specimens and living patients, he urged the students to attend clinical sessions at the Pennsylvania Hospital, the Philadelphia Almshouse, and the Jefferson College Dispensary.

As well as encouraging his students in the practical aspects of anatomy, Pattison excelled as a lecturer. His characteristics were later recorded by two of his students during his early years at the college, J. Marion Sims, the renowned New York gynaecologist, and Alfred Stillé, the American physician and pathologist. Sims recalled that when Pattison became enthusiastic 'he would forget himself and all around him, and would splutter and slobber and spit, the saliva flying in every direction, so that those who sat within a yard of him would be splattered all over. Of course the young gentlemen were too polite to say anything . . . and would watch, before he passed the amphitheatre, before raising their handkerchiefs to wipe it off.' In spite of this peculiarity, Sims was of the opinion that Pattison 'was the best lecturer on anatomy then living. . . . It made no odds what the subject was, the student was always chained to it as long as he chose to speak. We never tired of his enthusiasm or his eloquence. . . . He was very kind to the students and always managed to help them out of their scrapes. He lent them money and patronized them in every way that he could. He was a father to the students and sympathized with them in all their efforts.'[1] In like vein, Stillé claimed that it was Pattison's reputation that induced him to pursue the study of anatomy: 'I shall never forget his lecture upon the skull in which he recited with admirable feeling the famous lines of Byron beginning, "Is this a place where a god might dwell?" Indeed, the charm of Pattison's lectures was his enthusiasm, tempered and guided by cultivation. His voice was flexible and sonorous, though not loud, and his manner intensely earnest, but never violent.'[2]

The 1832 *Announcement* contains the first public mention of Pattison's having the degree of M.D. The initials appeared in all subsequent announcements, and in many obituaries he bears the designation 'M.D. (Jefferson Medical College of Philadelphia)'. Yet he did not, in fact, receive any such degree from the Jefferson Medical College nor from the Universities of Glasgow, Maryland, or London. A reasonable conclusion is that he conferred it on himself early in his new appointment. Support for this theory is found in the minutes of the Jefferson Medical College, which recorded Pattison as 'Esquire' on 8 March 1832, but as 'M.D.' on 28 March 1833. In the United States the degree was perfectly consistent with his qualifications, and if he had received his medical training there instead of in Britain he would have been granted it as a matter of course. Indeed, without it Americans might have called in question his position as a professor of medicine.

It is well to recall that all the M.D. degree signified was successful attendance at classes for two sessions of four months duration followed by oral (but no written) examinations. These were minimal requirements, catering to 'those young gentlemen who engage in the profession for the purpose of obtaining an income'. The professors offered a free, voluntary supplementary course of lectures for a further two months, ending with written examinations to identify those students worthy of medals and certificates of honour. A few years later, written examinations were introduced for even the basic course, as both professors and students increasingly recognized the well-known imperfections of 'orals', such as inconsistency of grading, inordinate waste of time, and personal idiosyncrasies: 'The situation of the candidate is embarrassing; some become agitated and lose their self-possession, and are thus unable to do justice to themselves; others again, by appearing to be so, enlist the feelings of the examiners, and, by their address, elude scrutiny; while, however conscientious and upright in his intentions, the examiner must have the common sympathies of our nature and is therefore always liable to be suspected of favoritism or prejudice.'[3]

Although occupied with planning new buildings, reorganizing the anatomy department, and preparing lectures and orations, Pattison found time to publish his views on the urgent and topical subject of cholera* and to serve as editor of a new medical newspaper entitled *The Register and Library of Medical and Chirurgical Science*. As well as writing editorials and articles for this periodical, he produced searing reviews of books by his medical colleagues. For example, two works, *Physiologico-Pathological Observations on Follicular Gastro-Enteritis* by E. Geddings and *The Principles of Medicine, founded on the Structure and Functions of the Animal Organism* by Samuel Jackson, are lambasted in his customary pretentious language: 'the very essence of affectation . . . a mere ostentatious parade of learning . . . slang of charlatanry . . . incongruous and badly sustained tropes and figures . . . gaudy rhetorical flourishes . . . agony of desire and exertion to strike and surprise . . . tumid, tasteless and affected . . . groans under the weight of pleonasm and tautology . . . unbridled desire to show his command of words.'[4]

At the conclusion of his first academic year, Pattison married. His bride was Miss Mary Sharpe of Cheltenham, England, and the

*See Appendix 1.

ceremony was performed in New York by the Reverend Dr J. M. Wainwright on 1 June 1833. Little is known about Mary. According to Samuel Gross, she was a Scotch lady exceedingly proud of her husband and devoted to him: 'It was pleasant to hear her speak of him invariably as her "dear Granville".'[5] They had no children and Mary outlived her husband by many years.

This period was a happy and successful one for Pattison. The London fiasco was over and forgotten. Lately and happily married, he enjoyed the respect of his colleagues and students. His success as a teacher was reflected in the size of his class, which in a year had grown from 60 to 189 students. In addition to his many duties at the college, he carried on a general medical practice, mainly among his medical students. J. Marion Sims recalled the case of a fellow student ('a handsome young fellow from Alabama by the name of Lucas') who in 1834 became very ill from contact with an infected cadaver. When Sims found Lucas lying in bed *in extremis*, he sent for Pattison who was attending him. When Pattison had examined Lucas, Sims asked him what was wrong with his friend. Pattison replied, 'Why, he has the smallpox and he is going to die tonight. I thought you were acquainted with what was the matter with him.'[6] Lucas did indeed die that night to the great consternation and fear of his fellow students.

The period 1834–36 was spent in consolidating the changes and improvements at the college. The summer of 1836 is notable for the arrival of Robley Dunglison as professor of the institutes of medicine and medical jurisprudence. An Englishman of outstanding professional reputation, Dunglison had served as personal physician to three American presidents—Thomas Jefferson, James Madison, and James Monroe—and was called in consultation to a fourth, Andrew Jackson.

In accepting the chair, Dunglison was aware that he was moving to a medical school held in odium by the professors at the rival University of Pennsylvania. Nathaniel Chapman, in particular, would have nothing to do with the young upstart institution which had on its faculty his hated rival Pattison. But Dunglison in turn had a very low opinion of Chapman: 'He was hollow; made protestations, I fear, which he did not feel; and was by no means tender in his

Robley Dunglison, professor of the institutes of medicine and medical
jurisprudence at the Jefferson Medical College, Philadelphia. (Malloch Rare
Books Room, New York Academy of Medicine)

observations on others. . . . He had no judgement whatever; and I have been repeatedly ashamed of the figure he cut on the floor of the Philosophical Society. . . . Strange, that so great a weight should have hung upon so small a wire.'[7]

Dunglison and Pattison had known each other before they became colleagues, and Dunglison had received his commission of appointment enclosed in a friendly letter from Pattison. Nevertheless, one of his first acts on arrival was to condemn Pattison's exaggerated claims for the college in the annual *Announcements*. He felt that they went far beyond the truth and he speculated on Pattison's motives for such boasting: 'Dr. Pattison had the conviction that mankind are to be forced into beliefs, and that it is idle to be too delicate if the object is to make merit known. People—he urged—will not appreciate it unless it is over and over again, and strongly, placed before their attention.'[8] Perhaps as a result of Dunglison's criticism, the language of the *Announcements* became more temperate.

No sooner had this problem been resolved than Dunglison became involved in a quarrel between Pattison and George McClellan, who had originally been good friends. Both were opinionated and intransigent, and the relationship had gradually deteriorated. Pattison had lost his respect for McClellan. For his part, McClellan would insult Pattison behind his back and then, a few minutes later, put his arm around his neck, with expressions of friendship, calling him by his first name.

McClellan, seeking testimony unfavourable to Pattison, had written a letter to a Mr Jones who had formerly been a student of Pattison's in London and later a pupil-resident with Pattison and his wife in Philadelphia, but who had left, for reasons unknown, with feelings of great hostility towards Pattison.[9] Jones had replied with a claim which was highly offensive to Pattison's reputation and conduct, affirming 'that Dr. Pattison and his wife had lived together before they were married'. McClellan passed the letter to Dunglison's half-brother, at the time a student in both McClellan's and Pattison's classes at the Jefferson Medical College; from there it soon reached Dunglison. He was shocked that such a letter, containing a charge of depravity against a fellow professor, should have been passed on to a student in that professor's class. He therefore confronted McClellan at his home the very same evening, warning him 'that he could not answer for the consequences if the transaction were to come to the knowledge of Dr. Pattison'. McClellan replied that he had no fear of Pattison were he challenged to a duel, whereupon Dunglison observed 'that Dr. Pattison would probably not challenge him, but would shoot him, and the world would hold him justified under the circumstances'.[10] Dunglison's timely intervention averted a physical encounter between the two protagonists.

Incidents such as these, which involved both students and faculty at the college, provided Dunglison with the background for his book *The Medical Student*, a significant contemporary account of American medical education. Dealing with all stages of the students' development, it is characterized throughout by practical common sense. In general, Dunglison deplored the exorbitant demands made on students' time ('Each professor never fails to magnify his own [department] by counting the cost of time and labour which [the student] must be prepared to bestow.'); the preoccupation with trivia ('What a waste of time in directing the serious attention of the student to insignificant points, the recollection of which may be a good exercise for the memory but can be of little or no advantage in after life.'); the pompous erudition of the terminologists (*philopatridalgia* for home-sickness and *laryngo-tracheite-myxa-pyo-méningogène* for croup); the overemphasis of the rare condition at the expense of commonly encountered afflictions; and the tendency of the student to rush to new remedies if the first were not immediately successful ('He will soon learn that infinite mischief can be done in this manner and that more reliance has to be placed upon the recuperative powers of the system.'). Throughout the book, Dunglison reiterated his belief that the goals of medical education should include not only superior learning and abilities, but also gentleness, honesty, and compassion.

The Iowa Copper Mining Company

Honesty and compassion were conspicuously absent in the next episode in Pattison's life. The story of his association with the Iowa

Copper Mining Company is one of high hopes, broken promises, and eventual bankruptcy for at least one of the persons involved. The attempt to launch the venture by means of British speculative capital was hampered by slow communication—the steamship being the fastest means available—and by a severe economic depression in the United States and Britain.

During the winter of 1836/37, Pattison became a major shareholder of a newly-formed company based in Philadelphia, whose board was interested in acquiring property believed to be rich in copper at Mineral Point, Iowa County, Wisconsin.[11] This was a period of prosperity for both Britain and the United States, and prospects were bright. Accordingly, in the spring of 1837, the company purchased 500 acres of mining ground and 800 acres of woodland for the price of $100,000. Pattison owned 40 of the 200 outstanding shares; this reflected the value of his notes ($20,225), made payable over 24 months toward the purchase price. While the negotiations were in progress, he was requested by the board to proceed to England, there to attempt to raise capital in exchange for a one-half interest in the property of the company. The shareholders formally appropriated the sum of $3000 for his expenses. Before he set sail from New York on 8 April 1837, he was handed a bill of exchange for £100, and $1000 in cash. He requested, but did not receive in time, powers of attorney, notarized copies of the deed of conveyance, the declaration of trust, and other necessary documents. Meanwhile, the economic climate in both countries was deteriorating disastrously as each day passed.

Pattison arrived in Liverpool on 24 April 1837, after a 'very delightful passage of sixteen days'. He soon started to make the rounds of business contacts and to seek out interested speculators. The firm to which samples of drillings had been shipped from New Orleans confirmed 'the high reputation of the ore', and the managing director assured him that, in normal times, there would have been no difficulty in persuading British capitalists to purchase one half of the property and engage with the company in mining. 'At present, however, they seem to think there is no chance of inducing English capitalists to make investments *in any kind* of American property. They assure me that never in their recollection has there been such a time as the present and that the year 1825 bears no comparison to it. ... It is to be hoped that there will [soon] be *some* improvement in the money market and that capitalists will have got over the panic which at present overwhelms them and prevents them making investments in American property.'[12]

Pattison did his best to attempt to negotiate an agreement. 'In my conversations with the gentlemen I have seen, I have not stated to them that the object of my mission was to dispose of one half of the mine. On the contrary, I have said that I wish to gain information as to the best methods of smelting, rolling, etc. and that, if we could meet with a party or parties who were familiar with these subjects, we might be induced to sell them one half of the property to secure the information they possess.'

But from the moment of landing he was hampered by the lack of documents supporting his claims. No contract could be signed until the papers arrived from the company. His irritation increased with the arrival of one boat after another without a single line of communication from the company. By 6 June his patience was well-nigh exhausted. He wrote to the company stating that he was now ashamed to meet any of the prospective backers, having waited in vain for delivery of the papers.

He had received some promising enquiries during the weeks of waiting. In May he wrote to the company that a price of £40,000–50,000 might well be obtained for the half interest. In later negotiations the figure of £60,000 was discussed favourably with one British firm. The necessary papers finally arrived on 13 June but, to his chagrin, the power of attorney was drawn up in such a manner that the price of the half interest was set irrevocably at $100,000, at that time worth about £20,500. Pattison believed that such a discrepancy in price, coupled with the worsening economic climate in the United States, would scare off the prospective purchaser. He therefore requested the company to send another power of attorney in which no price was mentioned. Meanwhile the British firm had become suspicious and declined to do anything further until two of its members had visited Mineral Point to examine the mine at first hand; subject to a favourable report, the firm would then conclude the purchase of the half interest at £60,000. Perhaps fearing that the true value of the mine might be discovered, the company did not consent to the proposal, and Pattison returned to Philadelphia after a frustrating and unsuccessful mission.

The troubles of the Iowa Copper Mining Company were not yet over. During the next few years, Pattison, John Andrews (the treasurer), and others proceeded with the development of the property, including the construction of a copper-smelting furnace on an eighty-acre lot, but the venture, about which Pattison had spoken in such glowing terms in London, was doomed from the start. The principals had no metallurgical or mineralogical training nor had they checked the vendors' extravagant claims about the copper potential of the property. By 1840 Pattison was in dire financial straits through his involvement with the company; he assigned all his interests to Andrews by deed poll of 12 March 1840. In turn, Andrews notified his attorney that he was now the sole trustee of the company, 'Dr. Pattison having relinquished his trust'. Later in the year the company, mortgaged to the hilt, was put up for sale. The United States was still suffering from the depression, money was scarce, and no sale was concluded. The company was steadily going downhill and, by 1841, its banker, the Bank of the United States, was considering its dissolution. Andrews assigned all outstanding rights of the company to the bank on 6 July 1844. In later correspondence between various interested parties, it was noted that Pattison had not settled his debts and that in fact he probably never would, because he had declared bankruptcy. Parts of the property were sold by the bank in 1848 and 1858 for undisclosed sums, and the final sale of the residual assets was consummated in April 1859.

A poignant side issue involved the original co-owner and covendor of the property, John D. Ansley. When Pattison and Andrews purchased the lands in 1837, they had not been told that Ansley had previously mortgaged the property to a James Davis Jr of Boston—or so they later claimed. When informed of the mortgage, the company's new directors accused Ansley of deception and fraud and declined to honour their notes issued in payment for the property. Ansley denied the accusation, stating that he had told Pattison and Andrews about the mortgage and that he had proposed

that Davis be reimbursed using the promissory notes issued by
Pattison and his fellow directors. Whatever the truth of the matter,
Ansley was bankrupt in 1839. He went to England on his own to try
to sell the company but failed because it denied him access to the
necessary papers. From this point on, Ansley was continually
harassed both legally and financially until his death in 1845. In 1838
and 1841 he was incarcerated in debtors' jails for several months.
During his periods of freedom he returned to the company's lands
where he tried to steal the equipment and destroy much of the
timber. The company thereupon issued an injunction 'to oust and
eject trespassers', thereby denying Ansley further revenge. After
trying in vain to sue for unpaid debts, he died in poverty, beset by
creditors, in the full realization that he had been cheated by the
company.

These events led to a number of lawsuits, in all of which Pattison
was named as a defendant or respondent as one of the original
purchasers of the property. In each case he was exonerated because
he had divested himself of all his shares. Davis sued Pattison and his
fellow directors for his outstanding mortgage in Chancery Court in
February 1840, October 1845, and November 1858 (Pattison had
been dead for seven years at the time of this last action), and Ansley
sued them in February 1841 for failing to honour their notes. In
1846 an attorney gave as his opinion that 'there would not be five
cents paid on the dollar' in the event of a sale, as by then the
company's only material asset was some real estate in a remote part
of Wisconsin. Nevertheless, the officers of the Bank of the United
States tried to recover outstanding debts by dissolving the company
in 1848 and selling parts of the property in a mortgage sale in
Chancery, 'the objects and intentions of the Iowa Copper Mining
Company being now unattainable'.[13]

In addition to its other liabilities, the company had defaulted in
reimbursing Pattison for his expenses on his trip to England in 1837.
He had received about half of the originally appropriated sum of
$3000 and had been forced to borrow money from, and pay interest
to, his British friends to meet his later expenses. In December 1839
he wrote two strongly worded letters to Andrews, pressing for
payment of the outstanding balance plus interest at six percent from

March 1837; no record has been found regarding the outcome of this demand. The affair of the Iowa Copper Mining Company was an unfortunate episode; clearly, Pattison was willing to undertake speculative financial operations with apparent ease and confidence, if not with success.

Departure from Philadelphia

If the Iowa Copper Mining Company caused Pattison considerable anxiety, so too did developments at Jefferson Medical College. Legislative changes in 1838 had freed the college from any connection with the parent institution at Canonsburg. Its newly acquired independence called for a restructured board of trustees and this was now causing unrest among the faculty. Its founder, George McClellan, in particular was behaving indiscreetly both within and without the confines of the college. He showed contempt for all authority, called the board 'a parcel of politicians', and claimed that the college was 'going to the dogs'. Pattison and Dunglison opposed McClellan's criticisms and lack of tact. As dissension grew, the board of trustees asserted its authority and, on the basis of a special committee report, resolved in May 1839 that 'the present faculty be depurated'.[14] Each member, if he wished to remain, was required to write a letter requesting reappointment; all complied except for McClellan and his brother Samuel. McClellan was not immediately dismissed from the chair of surgery, as some members of the board continued to support him; they were finally overruled, and his connection with the college ended on 10 July 1839. For better or for worse, the trustees had sacrificed the founder of the college and one of America's most eminent surgeons in order to maintain their own prerogatives and dignity.

Pattison applied for reappointment to the chair of anatomy and was re-elected unanimously. But McClellan's departure had had a deleterious effect on student enrolment: more than a hundred left with him. The loyal remainder consolidated their ranks and drafted a memorial expressing their entire satisfaction with all members of the medical faculty. Nevertheless, Pattison was not happy about the serious reduction in the number of students, nor was he optimistic about the future growth of Philadelphia. He began to look for fresh pastures and became attracted by New York—a large city with a sizeable untapped source of medical students and an almost limitless supply of interesting clinical material. Pattison's colleagues had noted his frequent visits to New York in January 1841, and it came as no surprise when he tendered his resignation. His friend John Revere resigned at the same time.[15] Their stated purpose was to help in the organization of the newly formed medical school of New York University. Pattison's resignation was accepted by the board of trustees on 2 April 1841, and his successor was appointed four days later.

The reorganized faculty of the Jefferson Medical College for 1841/42, with Pattison's and Revere's names missing, was publicly announced on 1 March 1841; the resignations thus became known to all the students in the middle of the session, with disagreeable results. 'The young gentlemen did not sympathize with them in their withdrawal from the school . . . [and] they [Pattison and Revere] were occasionally ruffled; generally, however, more owing to morbid sensibility on their part than to any action of disrespect on the part of the students.'[16]

The students' annoyance over the two resignations led to an episode which showed Pattison and Revere in a poor light. A student, Benjamin Neal, had spoken disparagingly of Revere in the dissecting room. This was reported to Revere who, as dean of the faculty, immediately sent for Neal and reprimanded him. There-upon Neal went to Dunglison to complain of the harsh conduct of Revere. 'I [Dunglison] told him to take no notice of it; that we were very desirous that nothing but the greatest respect should be paid to our seceding colleagues and that he had done nothing wrong in expressing his feelings.' Weeks later, however, when Neal presented himself for graduation at the end of the session, Pattison refused to examine him on the ground that he had not paid for his dissecting ticket. Thereupon Revere declined to ballot for him because he had not been examined by Pattison. Dunglison considered this to be vindictive collusion. Accordingly, he arranged that balloting should be postponed until an hour before the time of the commencement, thus bypassing the faculty. Dunglison and a majority of his colleagues recommended directly to the board of trustees that Neal receive his degree. As they approached the board for this purpose, however, they found Pattison already there, addressing the board on the same matter. When Pattison had finished, Dunglison reviewed the episode, and the board, without any deliberation, unanimously approved Neal for his degree. Dunglison remarked, 'I confess I have never been able to find the slightest shadow of excuse for this attempted gross act of injustice which gave me a worse opinion of my two friends than I had ever had.'[17] So soured were their relations with their colleagues that Pattison left Philadelphia without paying Dunglison a call or even leaving a card at his house.

Pattison escaped from all this acrimony when he arrived in New York in June 1841. It was his last move. The next ten years were to be the most tranquil of his life.

Life and Work in New York

New York University, incorporated in 1831, established a medical school in 1841.[18] Among those who, in 1829, started a movement to establish the university along progressive lines were the Rev. Dr J. M. Wainwright (who conducted Pattison's wedding in 1833) and Valentine Mott, one of Pattison's future colleagues. Pattison had already twice considered employment at the university before he was finally invited to be a founding member of the new medical faculty. After his dismissal from the University of London in 1831 he had offered himself without success as a candidate for the chair of anatomy, having solicited the support of his patron Albert Gallatin who was then head of a council for the establishment of a university in New York. And in June 1834 he received overtures from Mott, then professor of surgery in the College of Physicians and Surgeons in New York, about a proposed chair of anatomy at the university. Pattison had laid down such stringent conditions about the security of the appointment, his authority for controlling the dissecting rooms, and the necessity of increasing student fees for his lectures that again he was not offered employment.

Stuyvesant Institute, the first medical building of New York University.
(New York University Medical Center Library Archives)

Pattison's new colleagues were Mott, professor of surgery, the most famous surgeon of his day in America, about whom Sir Astley Cooper said that he had performed more of the great operations than any man living; Gunning S. Bedford, professor of midwifery and diseases of women and children, who was cruelly pilloried as 'the phenomenon' by the editor of *The New York Lancet*; Martyn Paine, professor of the institutes of medicine and materia medica, who was of the old 'lancet-and-calomel' school, and of whom it was said that to pass any of his examinations the only answer ever needed was 'the treatment is blood-letting, sir'; John William Draper, professor of chemistry; and, of course, Revere, who took up the professorship of the theory and practice of medicine. Pattison became professor of anatomy.

The six founding professors bought a large granite building known as the Stuyvesant Institute situated at 659 Broadway, the finest thoroughfare in New York. Lectures and clinics were held in the institute until September 1851, when the faculty moved to a larger, more handsome building which they erected on Fourteenth Street.

The medical faculty was required to be self-supporting, and the authorities were naturally eager to attract new students. Lectures commenced on the last Monday in October—one week earlier than in Philadelphia; this allowed prospective out-of-town students to attend the free introductory lectures in New York and enjoy a few days of city life. No doubt it was hoped that students might be lured away from the rival medical schools at the University of Pennsylvania and Jefferson Medical College. To allay the concern of students and parents about living accommodation, the first *Announcements* incorporated lists of respectable, inexpensive boarding-houses.

Fees for a full course of lectures amounted to $105; an additional $5 was required from those students wishing to avail themselves of the privilege of dissection, at that time not mandatory at New York University. To place these fees in perspective we can compare them with the standard charges made by doctors in the 1840s in the Eastern United States. At that time, office consultations cost the patient $1 to $10, with extra expense if the doctor had to travel over bad roads or at night or in inclement weather. Routine obstetrical care cost $10 to $30 per delivery; lithotomy (removal of a stone from the bladder), $50 to $200; the amputation of an arm or a leg, $25 to $100; the extraction of a tooth, $1; and the treatment of a case of gonorrhoea or syphilis ('fee in advance'), $10 to $50. (The high charge for treating venereal diseases reflects the disgrace felt by the patient, the contempt shown by the public, and the physician's risk of becoming infected at a time when bacteria were unrecognized and antibiotics unknown. Venereologists were usually extremely wealthy.)

The course of study was the same as that offered at Jefferson Medical College, that is, daily lectures in the standard six disciplines over a period of four months, with optional attendance in the dissecting room and at the surgical clinic. To graduate, students were required to attend two complete sessions, after which they were examined individually by the professors. In 1842/43 the faculty offered a two-month supplementary course for an additional $50, but later discontinued it as the students were disinclined to prolong the period of study.

This and much additional information was provided in the first *Announcement*, the style of which bears Pattison's unmistakable stamp—for example, in the boast that 'she [the new medical school] stands without rival in the United States'. Such propaganda yielded a first-year enrolment of 239 medical students, all Americans except for nine, who came from Canada, Demerara, and England. This figure rose to over 400 during the ten years that Pattison spent in New York. Indeed, wherever he taught, there was almost always a steady increase in enrolment.

In addition to his regular duties in the lecture hall and dissecting rooms, Pattison served as one of the two attending surgeons at the university's clinic, set up to relieve the bodily infirmities of the suffering poor and to attract more students to the new school.[19] All services were provided free of charge, and anyone able to pay was turned away. Pattison and Mott treated indigent patients for surgically correctable conditions and for diseases of the eye and ear. Enthusiastic audiences of over 150 doctors and students regularly attended the sessions, which were held every Saturday at noon for two hours; the curiosity of an even wider audience was served by the *New York Herald*, which provided extensive coverage of the new clinic's early months. Pattison and Mott were on good terms in these early days, joking and complimenting each other; they were soon to damage the reputation of the clinic by quarrelling publicly.

Pattison occasionally entertained the patients and audience with anecdotes about the lengths to which doctors were sometimes forced to go in order to collect their fees. One story concerned John Burns, his old friend and colleague in Glasgow, who was called to reduce the dislocated jaw of a very rich but miserly old gentleman. When the treatment had been successfully completed, the patient said, 'I like to encourage scientific surgeons—here, sir, is half-a-guinea [about one-twentieth of Burns's usual fee].' Burns thanked him profusely, engaged him in conversation for about half an hour and then gave some huge yawns. The quondam patient started to yawn in concert and off again went the ill-fated jaw! 'Now, sir,' said Burns with relish, 'I wish you a very good morning.' The patient made imploring gestures and stamped his foot. 'My fee for both operations,' said Burns, 'is twenty guineas.' This the reluctant patient was compelled to pay before Burns would make any attempt to relieve his discomfort.

Pattison was fifty years of age when finally he joined the faculty. A portrait painted during this period by an unknown artist shows a quizzically smiling, elderly man wearing a frock coat with tartan lapels. 'A small, elderly man, of medium stature, with black eyes and white hair', 'with graceful bearing, earnest manner and lisping voice',[20] he was considered kindly, affable and courteous. Sometimes penurious but never parsimonious, he was always among the first to respond generously to charitable appeals—'in his pecuniary relations he was often most injurious to himself'.[21]

In his dealings with his fellow professors he was said to be 'a zealous and popular colleague; a little too urgent, when a measure, perhaps of doubtful expediency, was embraced by him; but generally willing to yield to the voice of his colleagues if calmly and firmly expressed. . . . A great fault with Dr. Pattison was his unscrupulousness if a favourite object had to be carried; and the only way to manage him was to resist him, and if he found the majority were against him, he had to yield.'[22]

As a teacher, Pattison was invariably respected in New York by students and colleagues alike. 'As a lecturer on anatomy, [Dr Pattison] almost made the dead body before him speak, as with his own peculiar eloquence he gave a masterly exposition of the importance of anatomical knowledge as the only avenue to success and distinction. . . . His success in fortifying the student in the knowledge as well as with faith and boldness to *cut freely* when indicated—to *make free and bold incisions*—made him one of the most successful, as well as attractive lecturers that ever adorned a medical college.' The same student observed that he had 'an ardent love for medical students . . . and in return secured their strong attachment'.[23] A colleague observed that 'Pattison's forte as a teacher consisted in his knowledge of visceral and surgical anatomy and in the application of this knowledge to the diagnosis and treatment of diseases and of accidents, and to operations. . . . Pattison never indulged in any of those physical displays occasionally witnessed in our amphitheatres. On the contrary, he was dignified, entertaining and instructive.'[24]

Such eulogies were based more on Pattison's rhetorical skills in the extemporaneous presentation of his anatomy lectures than on their factual content, which amounted to no more than straight-

Granville Sharp Pattison, by an unknown artist, *c.* 1846. (New York
University Medical Center Library Archives)

forward accounts of the standard anatomical systems.[25] The feature of greatest interest to the students was his inclusion of a great many clinical applications.

Pattison had not lost his love of giving popular lectures to large and enthusiastic lay audiences. On one occasion, after a lecture-demonstration on 'the Human Frame', 200 young ladies crowded round him for half an hour, 'pulling all the bones about, looking closely at the different drawings of the human frame and asking the professor a thousand questions, all [of] which he answered naïvely and with great good humor'.[26]

Pattison's leisure activities while in New York included frequent visits to the opera and the concert hall. Although no performer himself, he was very fond of music. During the last year of his life, he was smitten by the great operatic diva Teresa Parodi, who made her North American debut at the Astor Place Opera House on 4 November 1850. 'No-one was more vociferous in his applause; and his attendance during the whole engagement of that celebrated singer was almost constant.'[27] If he were inclined, Pattison could have attended recitals by Jenny Lind and the seven-year-old Adelina Patti; appearances by Tom Thumb, who was two feet tall and weighed sixteen pounds; frequent productions of Shakespeare plays starring the Keans, the Booths, and William Macready, all of whom appeared in New York during his ten years there. He enjoyed partridge shooting, could cook a canvas-back duck to perfection, and was never happier than when casting a fly in a pleasant trout stream in May or June. But 'being naturally of an indolent disposition', he spent much time reading light literature, visiting his friends, and attending places of amusement. The *New York Herald* recorded his attendance at several important social functions, including the ball honouring the visit of the Prince Ferdinand de Joinville in 1841.[28]

Pattison the Surgeon

During Pattison's life, surgery was performed without the benefit of anaesthesia. Speed, dexterity, and an accurate knowledge of surgical anatomy, preferably accompanied by a calm and confident mien, were the prerequisites for the successful surgeon. Pattison possessed these attributes, although Gross reports that as he advanced in years the sight of blood became distressing to him. Yet, by today's standards, his attitude to his patients' pain seems insensitive and this in turn seems to have encouraged a certain callousness in his students.

On one occasion he was operating on the extra-ocular muscles of the eye of a young man with a very bad squint. After two or three cuts the patient roared out in agony, 'Stop, stop—only for a moment—let me rest a bit—take your hand off my eye.' 'There', said Pattison, calmly wiping his instrument, 'it's done—get up and wash your eye—and when you come here again just try and muster up a little more fortitude.' On another occasion, he was about to operate on a little girl with the same complaint, who had not been warned about the pain of an operation. She started to scream and struggle. 'Oh, my dear, it will not hurt you at all,' he said untruthfully, whereupon the students laughed and beat the floor with their canes. On yet another occasion, Pattison was using acupuncture to treat a young man for severe stammering. Before the second needle was quite through the tongue, the patient became angered and let forth a string of expletives. Again, the students roared with laughter. In an aside to the audience, Pattison remarked that in the course of a long practice he had invariably observed that females always bore severe surgical operations more stoically than men.[29]

In these days of anaesthesia, it is easy to forget the agony of patients under the surgeon's knife. Over the centuries, many concoctions of questionable value have been used for the relief of surgical pain, including alcohol, opium, hyssop, hashish, and extracts of the mandrake plant. Hypnotism was sometimes effective. Various disastrous physical methods were employed to induce unconsciousness, including prolonged compression of the carotid arteries, stunning with a mallet, and massive bleeding from an artery. Possibly the least traumatic procedure for diminishing the pain of amputation was compression of the nerves proximal to the site of surgery. Patients were tied down or held by as many as six assistants. Operations exceeding twenty minutes in duration often resulted in death from shock, euphemistically referred to as 'exhaustion of the nervous powers'. Operating rooms were usually located in the dome of a hospital so that the screams might not be heard. Some patients preferred suicide to surgery. Nor was patients' dread confined to the operation itself, for the after-treatment too was excruciatingly painful. Every flap of skin was excised instead of being reunited and, for compound fractures, dressings were thrust between the ends of broken bones. Because operations were reported by surgeons and not usually by patients, we hear more about the skill of the operator than about the suffering they inflicted.[30]

It was under such circumstances that Pattison performed a thigh amputation before an audience of 150 at the clinic on Saturday, 24 July 1841. The diagnosis was 'white swelling'.

> The patient was a youth about fifteen; pale, thin, but calm and firm. One professor [Mott] felt for the femoral artery, had the leg held up for a few moments to insure the saving of blood, the compress part of the tourniquet was placed upon the artery, and the leg held by an assistant; the white swelling was frightful; a little wine was given to the lad; he was pale but resolute; his father supported his head and left hand; a second professor [Pattison] took the long, glittering knife, felt for the bone, thrust in the knife carefully but rapidly—the boy screamed terribly—the tears ran down the father's cheeks—the first cut from the inside was completed and the bloody blade of the knife issued from the quivering wound—the blood gushed by the pint—the sight was sickening—the screams were terrific— the operator calm. Again the knife was thrust in under the bone; the terrific screaming was renewed; one or two picked up their hats to leave; scream on scream—and again the bloody blade of the knife issued from the wound and was laid aside. The flesh quivered and the boy cried agonizedly, 'Oh father! father! father! oh mercy! mercy!' The flesh was thrust back with a small piece of wet linen, the divided ends of the quivering muscles were sopped for blood with a sponge—the saw glistened in the hands of the operator—the father turned pale as death—the boy's eye fastened on the instrument with glazed agony—grate—crush—once—twice—and the useless

limb from the toes to the center of the thigh was quietly dropped into the tub under the table. At this moment, the father's eyes closed, his child's hand dropped from his grasp, he reeled from the table, and fell senseless on the floor.

The arteries were taken up, cold compresses only were applied—one or two stitches in the flesh—one or two more screams—and the boy was taken into an adjoining room and laid on a bed. The whole took less time to perform than the details have occupied in writing.[31]

By 14 August, the boy's flesh had healed by the first intention and the stump was clean and healthy. Finally, when he returned to the clinic on 11 September, Pattison commented to the audience, 'When this boy was placed on the operating table he seemed sinking from the effects of hectic fever. His pulse had been for a time as high as 120 in the minute and, in all human probability, death would soon have terminated at once his sufferings and his life. He is now in vigorous health—the stump has healed beautifully—and the lad is a living illustration of the blessings of scientific skill.'[32]

Pattison performed many other operations at the clinic during his first year in New York. Surgical procedures on the eye included the scarification of an inflamed eyelid, the partial excision of an eyelid, the removal of a pterygium, and the surgical correction of strabismus by dividing the extra-ocular muscles. He carried out this last procedure twenty-one times over a period of eight weeks with one hundred percent success; a reporter remarked that he had never witnessed this delicate operation performed with more dexterity, success, and steadiness of hand. It took a mere five seconds to complete, 'and the patients never even winced'.

Other surgical procedures varied in their complexity. Pattison drained abscesses, removed sebaceous cysts, excised warty tumours, treated deafness by passing a catheter into the Eustachian tube, ordered for a swollen and painful breast the application of forty leeches, circumcized a patient with congenital phimosis and ulceration of the glans, divided the tendons to relieve club-foot, performed skin grafts, and extirpated a vascular naevus from an infant by the repeated insertion of red-hot needles ('there was no haemorrhage and the child apparently suffered very little pain').

On a more heroic scale, Pattison performed extensive facial surgery on a fifty-year-old patient with a large tumour in the nostril. ('After the painful character of the operation had been explained, [the patient] manfully and without hesitation decided to have the operation performed.') An incision from the eye to the lip laid bare the nasal bone which was sawn down to its frontal attachment. The maxillary bone was then divided and a large portion removed using a pair of strong forceps. The tumour, which was seen to originate from the ethmoid bone, was removed by dissection. Finally, the nasal septum, damaged by the expansion of the tumour, was excised and other diseased tissue removed using a strong bistoury (a long, narrow, surgical knife) and cautery. ('The patient, a man of very firm, decided character, bore the operation most heroically.')[33] Reports during the following weeks indicated an uneventful recovery.

Pattison frequently employed acupuncture for the treatment of stammering, for which he passed three or four needles horizontally through the base of the tongue. Quite often, the patients, who on arrival could barely speak, afterwards thanked him fluently and with perfect ease. But Mott and Pattison, conceding that the good effects were often only temporary, sometimes resorted to dividing the geniohyoglossus muscle below the tongue.

As time passed, Pattison found the infliction of pain increasingly distasteful. Even the discovery of general anaesthesia, which pre-dated his death by a few years, did not diminish his recently acquired aversion for surgery. In his later years, he found lecturing on anatomy to medical students a gentler and more satisfying occupation.

The Mott-Houston Quarrel

The medical reporter for all of Pattison's surgical activities was James Alexander Houston, a vitriolic young Irish physician recently arrived in New York from his native land. He served first on the *New York Herald* before launching and editing *The New York Lancet*, which was also published by the *Herald's* publisher, James

Gordon Bennett. The weekly issues of *The New York Lancet*, which consisted of sixteen pages and cost three dollars a year, appeared regularly from 1 January 1842 until 4 February 1843, when, without explanation, publication ceased. Its motto was 'Vérité sans Peur'. During the periodical's brief existence, it was a source of much irritation to the medical faculty of New York University. With contrived insouciance Houston invariably misnamed the university medical school 'the Stuyvesant Institute Medical School', to the annoyance of the faculty and confusion of the public. Joshua Chamberlain commented, 'We are struck by the scurrilous, nay venomous tone of many of its chronicles, filled as they are with a bitterness of palpable ill-will, particularly towards Bedford and Pattison.'[34]

Soon after the start of the 1841/42 academic year, Houston became involved in a public quarrel with Valentine Mott, which culminated in a suit in the Court of Chancery. As a colleague of Mott, Pattison inevitably became involved. The publicity surrounding the affair provoked a subsidiary quarrel between Mott and Pattison. This all had a disastrous effect on the attendance at, and the reputation of, the university's clinic.

Initially, Houston was a staunch advocate of the new medical school. Throughout 1841, as we have seen, he gave laudatory reports in the *Herald* on the surgery performed by Pattison and Mott at the clinic, and reviewed very favourably the introductory lectures of all the medical professors. Houston's work so impressed Mott that he asked him to prepare verbatim reports of his lectures, at $5 per lecture, for future publication. At the same time, Mott asked Houston to guarantee in writing that he would not make public use of the notes. Houston refused but nevertheless continued to provide the regular reports on Mott's lectures.

Meanwhile, Houston had been appointed editor of *The New York Lancet*. Before the appearance of the first number, he offered to publish a synoptical review of the lectures. At first Mott concurred, requesting only that he see the proofs of the articles. Soon afterwards, however, he wrote to Houston forbidding him to take any further notes whatever at his lectures. Houston understood this to mean that the original agreement regarding the verbatim reports had been rescinded, not that he was barred from taking notes for his own purposes.

Houston published an account of Mott's first lecture in the first number of the *Lancet*. On the following Monday, Mott saw Houston at his lecture and confronted him before all the students. In addressing the class, he asked if they thought it fair that Houston should have accepted his money and yet was using the lecture reports for his own benefit. He felt that Houston was depriving him of his property.

Houston attempted to reply but was prevented by Mott with the remark, 'I'm Captain of the Deck here, if you please.' Mott then commented to the students that the publication of his lectures in the *Lancet* would 'take off all the freshness of the lectures when published by himself'.[35] Houston retired but attended Mott's lecture the following day. Again he was challenged ('What right have you here?'). In answer, Houston produced an admission ticket, which he had acquired in the usual way as a qualified medical practitioner wishing to attend university lectures. Mott shouted, 'I don't care for the order—you must go out—or I can't go on—that's all.' Houston again retired, but announced that he would be continuing to publish summaries of Mott's lectures.

A few days later, Houston was served with a bill of injunction, in which he was forbidden, under penalty of $10,000, from publishing any reports on Mott's lectures in the *Lancet*. He was further ordered to appear in the Court of Chancery on 12 January to answer a bill of complaint by Mott. The third number of the *Lancet*, due on 15 January, was suppressed until the case had been heard, and Houston made his views known through its sister paper on 10 January: 'The Captain of the Deck claims his lectures to be his exclusive property. They are no such thing. . . . His students can take verbatim copies and sell, print, publish or dispose of them as they please.'

His viewpoint was upheld by the Court: in a closely argued opinion, Murray Hoffman, the Assistant Vice-Chancellor, ruled in Houston's favour on condition that he keep an account of the profits of the *Lancet*. The ruling was announced in the *Herald* of 20 January 1842 under the headline 'Liberty of the Press Triumphant—Injunction on the Lancet Dissolved'.

Houston took revenge on Mott by levelling incessant criticism at the faculty and by publishing a scathing review in both the *Herald* and the *Lancet* of a recently published *Book of Travels* by Mott. The barrage of abuse finally induced Mott to insert a postscript in the 1842/43 Faculty *Announcement*, condemning Houston's attempts to sow discord among the medical students, his misrepresentation of the condition of the new medical school, and his perverted reports of medical lectures. Mott concluded the indictment by stating that Houston had been expelled from the medical college by a unani..ous vote of the faculty.

The quarrel spread to involve the clinic, but even before the suit of injunction was launched, jealousies had apparently been developing between Mott and Pattison. We have only Houston's accounts of these: how much was scandal unscrupulously worked up for his own purposes is not clear.

> Dr. Pattison is a most thorough anatomist—a judicious physician and a good surgeon; not so firm as Dr. Mott, but more scientific; not so ambidextrous, but more philosophic; not so witty, but equally good humored; not so rich as Dr. Mott, but equally celebrated in the domains of 'Surgery and Surgical and Pathological Anatomy'; Dr. Mott is sixty-eight, Dr. Pattison is twelve years younger—both of the old gentlemen, however, are in a fine state of preservation and wear their own hair.
>
> The Captain of the Deck was now told that his rival Dr. Pattison was rapidly getting into lucrative practice and that the *Herald* reports of the *clinique* were instrumental in producing this alarming result. Dr. Pattison heard this too, and immediately requested our reporter in future to omit the names, which was accordingly done.
>
> But it was intimated to Dr. Mott, a few weeks afterwards . . . that Dr. Pattison was getting a great deal of the merit of remarks and operations at the *clinique*, which by right belonged to Dr. Mott. The Captain of the Deck then requested our reporter 'not to be mealy-mouthed about the thing, but give the name of the operator in each case'. Our reporter complied.[36]

Mott's irritation peaked when he introduced at the clinic a young protégé, who performed some minor surgical procedures which the

Valentine Mott, professor of surgery at New York University, engraved by
W. G. Jackman after a photograph by Brady. (National Library of
Medicine, Bethesda, Maryland)

Herald failed to report. Incensed, Mott ordered the *Herald* to apologize for this 'unpardonable dereliction' and to puff his young protégé in the next issue. The order was naturally ignored, whereupon Mott tried to preclude any further reports of the clinic. When this too was unsuccessful, he summarily resigned on 8 January 1842. Ten days later Pattison resigned too, in deference to Mott. 'So the breaking up of the *clinique* and the ultimate ruin of the New School results from petty jealousies and despicable feelings of rivalry among the Members of the Faculty.'[37]

Houston reported the cancelled clinic of Saturday, 8 January under the headline 'The University Surgical Clinique Broken Up— The Medical Revolution in all its Glory'. About 300 students and medical men had attended at the usual hour, but the doors were closed. 'So the cripples and the crooked, the cross-eyed and the tongue-tied had to go their ways . . . without sympathy or succor.' Houston's satirical pen later described how the faculty might have explained the closure to the students, who were now cheated of attendance at the clinic: 'Dear young gentlemen, lovely first-born of an illustrious mother, do keep quiet; next year, come back, and we'll have an hospital for you, but—ahem!—"bring the needful".'[38] In an attempt to minimize the damage, the faculty gave a party for the medical students in March 1842, featuring roast goose, champagne, and cheers led by 'the wild, "hip, hip, hurrah" of the venerable Pattison'.

The clinic remained closed—to the great benefit of the rival Crosby Street Medical School (College of Physicians and Surgeons) —until 2 April 1842. It had been agreed in the meantime that Pattison would be in sole charge during the summer and Mott during the winter. 'Each will thus be enabled to do what is right in his own eyes, having none to make him afraid, no-one of whom to be jealous, and nobody with whom to quarrel. We are sorry however that Dr. Pattison will not have the management of the *clinique* when the students are in town, as he is certainly the most fluent and intelligible speaker, and it is a pity that he should waste his scientific sweetness on the close sepulchral air of the deserted amphitheatre.'[39]

To the embarrassment of the faculty, the dissensions received international publicity. The journal *La Lancette Française* reported in February 1842 on 'la jalousie de M. Valentin Mott . . . contre son jeune confrère M. Pattison'. Referring to the temporary suspension of the clinic, the report contained a comment that 'M. Pattison, cause première et innocente de ces graves débats, se trouve privé de la possibilité de continuer ses doctes leçons'.[40] Houston quoted this French report in *The New York Lancet*, and added that 'the family squabble . . . has peacefully terminated, by the magnanimous act of Dr. Mott, who had retired from the field and left it in the sole and undisputed possession of his *"jeune!* confrère M. Pattison", who now delivers "ses doctes leçons" with great energy and *éclat*, to asthmatic old women, ricketty children, and empty benches. *Vive la bagatelle!*'[41]

For the rest of the spring and summer of 1842, Houston repeatedly commented on the very poor attendance at Pattison's clinic. 'Dr. Pattison's *clinique*, we are sorry to say, was miserably attended. A child with sore eyes and an old sailor with spinal disease were the principal cases.' 'Dr. Pattison's *clinique*, as usual, was poorly attended. There were, however, a good many old women and ricketty babies present, to whom the learned Professor lectured with his usual energy and good sense; he also scarified an inflamed eyelid with remarkable dexterity and *steadiness of hand* [Houston's italics], and received the wondering admiration of the aforesaid venerable matrons and interesting juveniles.'[42]

The medical faculty suffered another setback, to which Houston was quick to refer. All their students who sat the examination for admission to the naval medical corps were failed by the board of examiners. He cited this as proof of 'the miserably inefficient system of education at the Stuyvesant Institute Medical School'. He suggested that circulars for the school should read: 'Students qualified to pass the green-room of the Stuyvesant, but not the Navy or Army Boards.'[43] The failure of these students bolstered Houston's opposition to what he perceived as the conservative stance taken by Pattison and his colleagues toward the reform of medical education. Houston favoured a more extended and liberal course of instruction, along the lines of that offered in Europe. Pattison, who, twelve years previously, had suffered at the hands of Thomas Wakley in the London *Lancet*, was now continually harassed by the editor of its

New York counterpart. As well as deriding the faculty for their 'cold-blooded, sneaking, crawling animosity' to educational change, Houston at times turned to personal invective:

> We observed the other day in one of our leading journals some sarcastic allusions to what it was pleased, in a somewhat unfeeling manner, to term 'the past migratory career of Dr. Granville Sharp Pattison'. Now, it may be very true that this gentleman has figured at sundry times, and in diverse manner, in some half dozen cities on both sides of the Atlantic, but we are not sure that any inference to his disadvantage must necessarily follow from that fact. On the contrary, such a career exhibits a remarkable illustration of indomitable perseverance and unconquerable hardihood. Many men after the reception of that unpleasant *shampooing* which is administered *a posteriori*, would anxiously seek 'a lodge in some vast wilderness', and gain the friendly covert of 'some boundless contiguity of shade'; but here is the astonishing spectacle of a man who manifests a stoical indifference to buffets—makes, whenever required, a polite exit—and when driven from one city— flees unabashed to another![44]

The Last Years

The unexpected and unexplained discontinuation of *The New York Lancet* early in 1843 spared Pattison further annoyance and embarrassment from Houston's pen. Indeed, from this point on, his life became much less eventful, as he accepted the role of elderly, respected professor, opposed to reform in medical education. Only in his domestic life did he espouse change: during their ten years in New York, he and his wife lived at no fewer than seven different addresses.

In this last period of his life, Pattison was active in advising American colleagues about candidates for forthcoming university appointments. Dunglison, for instance, corresponded with him regarding a replacement for John Revere, who died in 1847 from epidemic typhus. And in June 1850 Pattison volunteered his advice on the replacement for the chancellor of New York University. His letter is of particular interest for several reasons. It displayed remarkable tact and a gift of clear and cogent argument and it touched directly on the abolition of slavery, already a highly inflammatory issue in the United States. Through his christening, Pattison had been associated with an abolitionist of an earlier generation, Granville Sharp; by a stroke of irony it is precisely with names that he was now concerned. One of the leading candidates for the position of chancellor was a certain Reverend Mr Tappan who ran a young ladies' school in Bleecker Street. But 'Tappan' was also the name of a well-known abolitionist of that day. In opposing the candidacy, Pattison was careful to set aside the moral aspect of slavery itself, which could only have complicated the issue, and to concentrate on the inevitable effect of Mr Tappan's election on the welfare of the medical faculty.

> Independently of every other consideration, Mr. Tappan's *name* is one which, as Chancellor, would ruin the Medical Faculty. I enter not on the question of slavery. Two facts must be admitted. 1st. The South is feelingly alive, and more particularly so at the present moment, in reference to abolition. 2nd. Nearly one half of the Medical Class is derived from the slaveholding States. If the announcement goes forth that the Council has elected *Mr. Tappan* their Chancellor, unless they at the same time announce that *their* Mr. Tappan is not *the* Mr. Tappan, the Great Apostle of Abolition, we most assuredly will never receive a medical student south of Mason's and Dixon's Line. The very great, I may say the unheard of, progress and prosperity of the Medical Department of the University has naturally aroused the jealousies of rival Institutions. If they have the power to circulate a report that we are an Abolition School, all the energy and zeal of the Medical Faculty would fail to sustain it.[45]

Mr Tappan was not elected.

By now, Pattison's health was failing. Repeated attacks of cholecystitis led to his death on 12 November 1851, at the age of sixty. In a letter of the same date, Meredith Clymer, the pioneer neurologist and recently appointed professor of the institutes and practice of medicine, informed Dunglison of the news:

> As an old associate of Dr. Pattison, I hasten to inform you of his death which took place this morning. Before leaving for Europe, I attended him in a severe attack of what I supposed to be hepatic colic. I left him quite ill when I sailed, and on my return, found him very much changed. Every day he seemed to fall off, and last Wednesday I was sent for, and found him in great agony from a similar attack to that of last May. On Monday morning he began to sink, and died early this morning. On examination we found a number of gallstones in the duodenum and common duct, one of which, in the latter, had caused ulceration and had allowed bile to escape into the peritoneal cavity. What was most singular, that although a good deal of recent bile was found there, there was no recent peritonitis. I feel that, as an old colleague of the doctor, you will probably take some interest in these details and am sure that, with myself, you will regret the loss that we have sustained. My intercourse with him, since my connection with the school, has been of the most agreeable kind. I have always found him kind, courteous, and conciliatory. Under severe suffering, he was most patient, even-tempered, and confiding.[46]

His death was announced the same day to the medical faculty by the president of the university. At the meeting, his colleagues resolved that, in honour of his memory, they would wear black crape armbands for thirty days. His funeral service at 12.30 p.m. on 13 November in New York's Protestant Episcopal Church of the Ascension preceded burial in Greenwood Cemetery. The following spring, his body was moved to his homeland for reinterment in the Pattison lair of the Glasgow Necropolis, where he now lies in company with his parents and siblings.[47]

Pattison's will listed his personal assets: a one-sixth interest in the newly completed medical school building on Fourteenth Street, amounting to $5000; a $5000 policy with the Albion Life Insurance Company; and a $1000 policy with the Mutual Life Insurance

Company. ($11,000 in 1850 would be equivalent to about $131,000 today.) Testamentary disposition involved, after the payment of outstanding debts, the bequest of three quarters of the balance to his widow Mary, in trust, and one quarter to his sister Margaret, also in trust; both ladies received the income from the investment of their bequests but had no control over the capital. He further directed that, when Mary or Margaret died, the survivor receive the whole estate, still in trust. Finally, the survivor was empowered to dispose of the estate to whomsoever she wished, by drawing up an appropriate will. Two persons *not* named in the will were Pattison's putative daughter and her mother, Mrs Andrew Ure.[48] After Pattison's death, Mary returned to Britain where she was granted from the widows' fund of the Faculty of Physicians and Surgeons of Glasgow an annuity of £45 per annum; this was later increased to £52.5.0. The date of her death is not known; she was in good health as late as 12 May 1869.

Anatomist and Antagonist

Pattison's life reflected many of the sociological and educational issues of the day. As an outstanding teacher, he had occupied five chairs of anatomy, including the inaugural chairs at London and New York, and founded one of the first modern residential teaching hospitals in the United States. Yet he had seldom been free from controversy. In his frequent clashes with authority, he had been accused of body-snatching, adultery, and malpractice. He had quarrelled publicly with senior colleagues and had been eager and willing to challenge his opponents to duels. In most of these episodes he had been the aggressor; in London the roles were reversed when his students, perceiving him to be incompetent, had succeeded in having him dismissed.

Such a tumultuous life can be traced back to events in Pattison's early years. As a boy he was subjected to a variety of social pressures and indignities. Foremost among these was the family's fall from a life of affluence and luxury to a modest, middle-class existence. The former spawned a certain arrogance and slothfulness, while the latter forced on him the ambition to excel in his career as an anatomist.

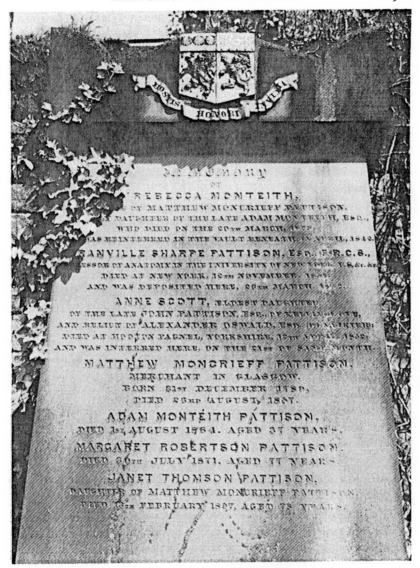

Granville Sharp Pattison's final resting place in the Glasgow Necropolis.
(Note the spelling error in his name.) Photo by Tom Steven.

He was undoubtedly a poor student at the Glasgow Grammar School, obstinate in his refusal to conform. Later, he seemed almost to court violence; when crossed, he became truculent and aggressive. These tendencies were augmented by frequent crises and scandals, the effects of which he appeared to offset by affecting a vain and conceited bearing. Nevertheless, throughout his life he was acknowledged by students, colleagues and friends to be invariably generous, fair and kind. He enjoyed attending the opera and other fashionable events—for which he dressed impeccably—yet he was equally happy in outdoor activities. Politics and religion played only minor roles in his life.

There is no question that his extraordinarily colourful and interesting life illuminates the contemporary scene, yet he cannot be described as one of the major innovators in British or American nineteenth-century medicine. His most significant publications were in the field of surgery, which included abdominal and vascular operations, removal of osseous tumours, and lithotomy. The validity of some of these was challenged, and indeed formed the basis for public exchanges and a pamphlet war. Less contentious was his revised and enlarged edition (1823) of Allan Burns's *Surgical Anatomy of the Head and Neck*, in which he provided many additional cases and observations. In his attitude to patients, he demonstrated a certain callousness and indifference to suffering.

It is as an anatomist and teacher that he will be remembered. His interest and enthusiasm were originally sparked by the teaching of James Jeffray and Allan Burns in Glasgow. He was soon lecturing to large classes of eager Scottish students at the College Street Medical School. Throughout his life he was invariably considerate and supportive towards his students, and in return (with the notable exception of the students in London) he secured their loyalty and strong attachment. He was widely considered—particularly in the United States—the best lecturer in anatomy then living, and anatomy was the single most important subject for medical practice at that time. Indeed, his main influence on medicine was through his students, whom he inspired with his enthusiasm and eloquence. He was never happier then when instructing in the dissecting room or when lecturing to large and appreciative audiences, both lay and professional.

With characteristic gusto he attacked the two main impediments to the training of medical students in his day—in Britain, the dearth of cadavers for dissection, and in the United States, the lack of clinical teaching in a hospital setting—and made significant contributions to the resolution of both problems. His spirited testimony before a House of Commons committee in 1828 carried much weight and was certainly an important factor in the eventual passage of the Anatomy Act in 1832; thereafter, subjects for dissection became readily available from anyone having lawful possession of a deceased person. His instrumental role in the founding of the Baltimore Infirmary, a forerunner of today's teaching hospitals, was his most important contribution to medicine and medical education in the United States. The provision of clinical training and experience was unique at the time; nearly three decades later, only a quarter of American medical schools required students to attend hospital at all.

It was perhaps inevitable that his outstanding career and flamboyant personality should excite both fervent admiration and fierce antagonism, an explosive combination which was a hallmark throughout most of his life. It is tempting to see in Mary, his wife, a stabilizing influence, to believe that their union was a watershed between the turbulence of Pattison's earlier life and the relative tranquillity of his last years. In the year of their marriage, Pattison published a lecture reflecting his newly-acquired philosophy of life. Characteristically, he shared this with his students. These were his closing words[49]:

> Many of you, gentlemen, are now about to enter into the world: let me, I pray you, impress on your minds a lesson I have derived from my own experience; which is, that if you have only, by diligence and attention to your studies, qualified yourselves for the performance of the duties of your profession, neither slander nor misrepresentation can prevent your success. When, therefore, you hear that you have been traduced, when your skill and ability are belied or called in question, do not, my young friends, allow your tempers to be irritated; do not repel calumny by ill-tempered recrimination;

recall to mind the axiom delivered by the Roman—which is not more remarkable for its truth, than for the consolation it must afford to the deserving—

MAGNA EST VERITAS, ET PREVALEBIT.

Appendix 1
Pattison's Publications

Glasgow (1815–19)

'Observations on abdominal wounds with cases, etc.' reported by
L.-R. Villermé in *Journal Universel des Sciences Médicales* 13
(1819): 241–52, and by an anonymous writer in *Journal Général
de Médecine* 66 (1819): 388–96.

Philadelphia and Baltimore (1819–26)

*Syllabus of a Popular Course of Lectures on General Anatomy and
Physiology, as illustrative of the natural history of man* (Phila-
delphia: privately printed, 12 October 1819).
'Experimental Observations on the Operation of Lithotomy, with
the description of a fascia of the prostate gland which appears
to explain anatomically the cause of urinal infiltrations and
consequent death', *American Medical Recorder* 3 (1820): 1–24
and two plates.
*A Refutation of Certain Calumnies Published in a Pamphlet Entitled
'Correspondence between Mr. Granville Sharp Pattison and
Dr. Nathaniel Chapman'*, 28 November 1820 (Baltimore:
Robinson, 1820).

A Final Reply to the Numerous Slanders circulated by Nathaniel Chapman, M.D. (Baltimore: Robinson, c. November 1821).

'A Reply to Certain Oral and Written Criticisms, delivered against an Essay on Lithotomy, published in the January number of the American Medical Recorder', *American Medical Recorder* 3 (1820): 351–71 (later published as a pamphlet [Philadelphia: 1820]).

An Answer to a Pamphlet entitled: 'Strictures on Mr. Pattison's reply to certain Oral and Written Criticisms, by W. Gibson, M.D.' (Baltimore: Matchett, 1820).

'Two cases of aneurism, with observations', *American Medical Recorder* 3 (1820): 193–202.

'Review of "An Essay on the Nature and Treatment of that State of Disorder generally called Dropsy" by J. G. Whilldin', *American Medical Recorder* 3 (1820): 590–605. Attributed to Pattison by Henry, *Standard History*.

'A case of Anastomosing Aneurism of the Internal Maxillary Artery, successfully treated by tying the common carotid artery', *American Medical Recorder* 5 (1822): 108–15.

Editor of: Allan Burns, *Observations on the Surgical Anatomy of the Head and Neck, Illustrated by Cases and Engravings*, 2nd ed., with a life of the author and additional cases and observations by Granville Sharp Pattison (Glasgow: Wardlaw and Cunninghame; London: Longman, Hurst, Rees, Orme, Brown and Green, 1824). The first American edition was identical (Baltimore: Lucas, Coale, and Cushing and Jewett; Philadelphia: Carey and Lea, 1823).

Co-editor of *American Medical Recorder*, e.g., 3 (1820).

London (1826–31)

Statements respecting the University of London, prepared at the desire of the Council, by nine of the professors (London: Spottiswoode, May 1830). (Pattison was one of the nine professors.)

Observations on a letter addressed by Leonard Horner, Esq. to the Council of the University (London: Spottiswoode, June 1830), pp.3–16.

Testimonials transmitted to the Council of the University of London (London: Spottiswoode, 1830).

Professor Pattison's Statement of the Facts of his Connexion with the University of London (London: Longman, Rees, Orme, Brown and Green, 1831).

Many letters and commentaries in newspapers and journals, notably *The Lancet*, as indicated in the Bibliography. See for example, *Lancet*, 1830–31, ii, 793–95, 825–29; 1831–32, i, 82–87, 209–15.

Philadelphia (1832–41)

A Discourse Delivered on Commencing the Lectures in Jefferson Medical College, Session MDCCCXXXII–III (Philadelphia: French and Perkins, 1832).

Letter on Cholera, from Professor Pattison to Dr. Carmichael of Virginia (Philadelphia: Geddes, 1832). The letter was originally published in the *Washington Telegraph*.

Pattison presented a long, historical account of the disease, which predated the identification of bacteria, and responded to three questions posed by a Dr Carmichael of Fredericksburg, Virginia. *Is cholera asphyxia contagious?* Pattison contended that it was not an infectious disease but was produced by a vitiated 'cholera atmosphere'. *What causes operate in its production?* He believed that electrical, magnetic or other changes in the atmosphere were responsible for the disease. *What system of treatment has been found the most successful?* He recommended heroic measures, including cupping, phlebotomy, emetics, morphine, castor oil, and massive doses of calomel. In a prophetic aside, he mentioned that George McClellan had had great success by the intravenous injection of seventy ounces of saline.

A Lecture Delivered in Jefferson Medical College, Philadelphia, on the question 'has the parotid gland ever been extirpated?' (Philadelphia: published by the students, 1833).

This is a witty and learned rebuttal of William Gibson's assertion that the parotid gland never had been, or ever could be surgically removed.

'Reviews', *Register and Library of Medical and Chirurgical Science*
1, no.45 (1834): 386–407. Editorial attribution.
*Professor Pattison's Introductory Lecture, Jefferson Medical College,
Session 1838–39* (Philadelphia: Waldie, 1838).

He praised Napoleon Bonaparte as 'the greatest man of
modern times' and quoted his declaration: 'To him that wills
and determines a thing shall be, nothing is impossible.' He then
eulogized John Hunter as 'The Father of British Surgery' and
Philip Syng Physick as 'The Father of American Surgery'.

Editor of *The Register and Library of Medical and Chirurgical
Science* (1833–36).

This 'medical newspaper', which ran for only three years,
has been described by W. D. Miles in the *Records of the
Columbia Historical Society* (1969–70): 114–25. Devised by
General Duff Green, an influential Washington political publisher during Andrew Jackson's presidency, the *Register*, at an
annual cost of $10, provided weekly issues of 64 pages devoted
to medical reports, articles, notices, and reviews. Its special
attraction was the regular serial publication of pirated reprints
of the latest and most authoritative European books, at a
fraction of their original cost. The last book to be issued in full
was Velpeau's *New Elements of Operative Surgery*, translated
by Pattison. Soon afterwards, in 1836, Duff Green abruptly
'killed' the *Register* without warning or apology. A few months
later, Pattison encouraged Robley Dunglison to launch a similar
publication, *The American Medical Library and Intelligencer*,
but was never actively involved in its production.

Editor of: Alf. A. L. M. Velpeau, *New Elements of Operative
Surgery, with an Atlas of nearly Three Hundred Engravings,
and an Appendix of Notes by Granville Sharp Pattison*
(Washington: Green, 1835).

The engravings and appendix have not survived, if indeed
they ever existed. Subsequent editions, edited by Valentine
Mott, did have an atlas in quarto of twenty-two plates.

New York (1841–51)

'Removal of a Carcinomatous Tumor from the Nostril', *New York Lancet* 1 (1842): 12–13.

Editor of: J. Cruveilhier, *The Anatomy of the Human Body* (New York: Harper, 1844).

The first American edition was prepared from the second Paris edition. In the Preface, Pattison declared that, so long as he was a teacher, he himself would never publish an original volume on anatomy because any book based on his own lectures would make it 'much more difficult to fix the attention of the pupils'.

Editor of: J. N. Masse, *A Pocket Atlas of the Descriptive Anatomy of the Human Body* (New York: Harper, 1855). Pattison's preface is dated 20 October 1845.

Introductory Lecture Delivered on the Commencement of the Session of Lectures of 1850–51, University of New York (New York: Gray, 1850).

Pattison acknowledged in this, his last published lecture, that by now his hair was silvered by the frosts of many winters. In the last paragraph, he coined the word 'Go-aheaditiveness' to encompass energy, incessant activity and great achievements. 'May this spirit endure for ever! Amen, amen.'

Appendix 2
Pattison's Family

Wise folk, daft folk, and they Pattisons![*]

GRANVILLE SHARP PATTISON's immediate family comprised his parents (John Pattison of Kelvingrove and Hope Margaret Pattison, née Moncrieff) and their eight children (in order of birth: John, Alexander, Anne Scott, Matthew Moncrieff, Frederick Hope, Granville Sharp, Margaret Robertson, and Hope Margaret). I have selected for inclusion in this Appendix only those who are known to have been actively involved in Pattison's life; their names appear in the Index. In addition, I have included for general interest very brief notes about another brother, Frederick Hope, because of his association with the Duke of Wellington at the battle of Waterloo, and an uncle, Alexander Pattison, because of his association with Robert Burns. For some of the information that follows, patchy and unbalanced as it is, I have drawn on family notes; these and other sources are listed in the bibliography.

[*] This is an old Glasgow saying.

John Pattison of Kelvingrove (1747–1807), father

Granville's father, born in Paisley, moved to Glasgow and established himself as one of the most successful manufacturers of muslin. The fine cotton yarn required for this was spun by machinery first introduced to Glasgow by John Pattison at the dawn of the Industrial Revolution. Weavers then produced the muslin on looms in their homes; at times Pattison had as many as 700 working for him.

He married Hope Margaret Moncrieff, from one of the most respected, ancient families of Scotland. He and his wife both had their portraits painted by Sir Henry Raeburn. John Pattison was armigerous, *bearing* argent gutté de sang a lyon rampant sable gutté d'or, on a chief azure three escallops argent. *Crest:* a camel's head and neck sable gutté d'or, gorged with a plain collar and crowned with an antique crown of the last issuant from a ducal coronet of the same. *Motto: hostis honori invidia.*

The Pattisons lived at Kelvingrove House, a mansion designed by Robert Adam surrounded by twenty-eight acres of wooded land. John Pattison's extensive improvements to Kelvingrove put a great strain on the family finances, and his business affairs also deteriorated during the last years of his life. He railed against exorbitant taxes imposed by the fiat of the Privy Council in London and even thought with envy of those who had fled the country. Eventually he was forced to sell Kelvingrove in 1806 and move with his family to a terraced house in Carlton Place, overlooking the River Clyde.

Politically, John Pattison espoused traditional Liberal or Reform sentiments. At times his effigy was exhibited on the lamp-posts in the Trongate of Glasgow during the day and burned in the evening as a democrat. On at least one occasion he was 'watched' by a government agent. An anonymous publication of the time, *Asmodeus; or Strictures on the Glasgow Democrats,* dubbed him 'Nimrod Heddles'.

The family worshipped at the Ramshorn Church in Canon Street (now Ingram Street), a mainstream member of the 'Auld Kirk' (the Church of Scotland).

Pattison died, aged 60, from a fever on 28 December 1807, the

Hope Margaret Pattison, née Moncrieff, by Chester Harding, 1825. From the author's portrait collection. Photo by Gary Mulcahey Photography.

year after he sold Kelvingrove. He was buried in the Ramshorn Churchyard. His dying wish was to be buried in a single grave with his wife, and on her death in 1833 three of his sons transferred his remains to the Glasgow Necropolis.

No record exists regarding his assets or their value at the time of his death; earlier the poet Robert Burns had estimated Pattison's estate in 1788 to be worth some £20,000. In his will, dated 25 March 1801, he bequeathed a life annuity of £350 to his wife (subject to forfeiture if she should remarry), a lump sum of £500 to his eldest son, John, and the residue in equal shares to his eight children. Three codicils appended to the will show the testator's growing confidence in his eldest son, who eventually assumed sole trusteeship under the inspection and control of his mother.

Hope Margaret Pattison, née Moncrieff (1755–1833), mother

Granville's mother was the daughter of the Reverend Matthew Moncrieff of Culfargie, whose lineage can be traced back to Somers Moncrieff in the twelfth century. It was through her, too, that Granville could claim kinship with John Clerk—later Lord Eldin— who defended him so ably in the grave-robbing trial and who provided a letter of support for his candidacy for the chair of anatomy at the University of London. Granville had two medical forebears through his maternal line: his great grandfather, Dr John Scott of Coates, and John Scott's father Robert.

Hope can rarely have been free of anxiety. Her husband's politics brought him under continual government suspicion and scrutiny; two of her sons, Alexander and Frederick, were in the thick of the fighting abroad; Granville was seldom out of trouble; and the family's business affairs were sometimes precarious. Her portrait was painted not only by Raeburn but also by the fashionable American artist Chester Harding during his visit to Scotland. She died at the age of 78.

John Pattison (1783–1867), brother

John, the eldest of the eight Pattison children, behaved in a responsible and generous manner on the death of his father in 1807. He bought commissions in the army for Alexander and Frederick, and financed Granville's medical education. In Philadelphia he provided help, encouragement, and hospitality for Granville during the trying winter of 1819/20. After spending many years there as a merchant, he returned to Glasgow, where he lived comfortably as a respected citizen. A proud man, he was one of the first in Glasgow to own a carriage and horses, for which he used a postillion. His pride impelled him on one occasion to send his son Alexander Hope (1812–34) to Ireland to fight a duel over a point of honour.

His portrait, also painted by Chester Harding, shows a striking resemblance to Napoleon Bonaparte, which led to an episode recorded by his grandson, also Alexander Hope (1843–1914), in his memoirs: 'He was the exact likeness of the great Napoleon in face and figure. When in London looking on at a review of troops, one of the generals thought he must be Napoleon and asked him to inspect the regiment. He carried out the joke by doing so, having been a volunteer officer. *The Times* next day had a long article about a "prince Napoleon" being in London.'

As well as resembling Napoleon, on one occasion John actually saw him. On 7 March 1803, during the lull between the Peace of Amiens and the resumption of hostilities, John wrote to his mother from Paris: 'I have at last seen Bonaparte, the conqueror of Italy, the great pacificator, the god of the French. There is something either in his face or in my eyes that makes him look above the rest of mankind. He is little in stature and pale, but his eye, tho' sunk, is the eye of the eagle. . . . He was mounted on his favourite charger which he rode at the battle of Marengo. . . . After reviewing about ten thousand men (far inferior to ours in discipline as well as looks), he returned to his palace not to be seen for another month.'

When John was twelve years old, he met Robert Burns. Riding through Dumfries, his father hailed Burns, who was standing on the steps of the inn. John later wrote to the *Glasgow Citizen* (on 5 January 1848) of the meeting: 'He who had remained motionless

John Pattison, by Chester Harding, 1825. (Glasgow Art Gallery and Museum)

till now, rushed down the steps and caught my father by the hand, saying "Mr Pattison, I am delighted to see you here; how do you do?" I need not say this was our immortal bard.' The three, joined by a friend, had dinner together from four till midnight. John was permitted to stay up for the occasion. 'I can never forget the animation and glorious intelligence of his countenance, the rich, deep tones of his musical voice and those matchless eyes, which absolutely appeared to flash fire and stream forth rays of living light. It was not conversation I heard; it was an outburst of noble sentiment, brilliant wit, and a flood of sympathy and good-will to fellow-men.'

John had eight children, including a daughter, Isabella, whose sad romance provides a poignant vignette of Victorian mores. She was in love with an officer in the Indian Army. On the day that he was due to sail, he offered no proposal of matrimony but instead g..ve her a pair of kid gloves. He sailed for India and was never heard from again. Meanwhile Isabella slowly pined away and died at the age of twenty-three of a 'broken heart' (officially 'nervous fever'). When her parents were arranging her belongings, they discovered the gloves and felt a hard object in the ring finger. This was found to be a gold engraved engagement ring. It must have been intended as a novel and exciting means of proposal. But Isabella, to keep the gloves in unsullied perfection, had apparently not tried them on and had not discovered the ring. Doubtless, the soldier, on hearing no word, supposed that he had been rejected, while Isabella, being unaware of the ring, thought that she had been abandoned.

The Pattisons, under John's leadership, seem to have had a taste for sepulchral splendours. Their burial ground in the Glasgow Necropolis is an impressive collection of family monuments dominated by that of his brother, Lieutenant-Colonel Alexander Hope, towering 25 feet. Granville's body joined those of his family when it was transferred there from New York in 1852, and John was buried there in 1867.

Frederick Hope Pattison (1789–1877), brother

Frederick served as senior lieutenant and company commander in the Duke of Wellington's own regiment (the 33rd) at the battle of Waterloo. Towards the end of his life—and some fifty-four years after the battle—he wrote a series of letters to his grandchildren which were published in 1873 for private circulation: *Personal Recollections of the Waterloo Campaign*. In this he offered a personal account of the battle, with a first-hand description of the Duke of Wellington himself and many anecdotes depicting the horrors of war in those days. From his own experiences, he concluded that one of the main causes of Napoleon's defeat was his failure to supervise and guard the French army's flanks. While the battle was raging, my maternal ancestor, whose namesake I am, Captain (later Rear-Admiral Sir) Frederick Lewis Maitland was waiting at the port of Rochefort on H.M.S. *Bellerophon*; less than a month later, Napoleon surrendered to him.

Margaret Robertson Pattison (1794–1871), sister

Margaret emerges as a gentle but steadfast figure in Granville's life. Living with her brother John in Philadelphia, she was able to welcome Granville on his arrival in 1819. Less pleasant was the public accusation made twelve years later in London by Thomas Wakley, editor of *The Lancet*, that Allan Burns's association with Granville was motivated more by Margaret's charm than by her brother's competence as an anatomist. In the end, Granville recognized her loving support by bequeathing one quarter of his estate to her.

Alexander Hope Pattison (1785–1835), brother

Alexander, while on furlough, served as Granville's second in the negotiations leading up to the duel with General Cadwalader in 1823. Earlier, he fought actively during the whole of the Peninsular War, was twenty-eight times under fire, twice wounded, and received a battlefield commendation at Salamanca. By 1830 he was a lieutenant-colonel in command of the Second West India Regiment in the Bahamas. His nephew and namesake joined him in Nassau as his aide-de-camp; both died of yellow fever during the winter of 1834/35. A touching monument of the twenty-one-year-old nephew, surmounted by accoutrements of war, stands near the impressive statue of his uncle in the Glasgow Necropolis.

Alexander Pattison, uncle

Alexander Pattison, John Pattison of Kelvingrove's brother, lived in Paisley. He attained modest renown as a friend of Robert Burns, whom he housed on the latter's visits to Paisley. Alexander promoted the sale of the first Edinburgh Edition (1787) of Burns's poems so successfully that Burns jokingly dubbed him 'Bookseller'. But Alexander is best remembered as the subject of an unflattering exchange of letters in February 1788 between Burns ('Sylvander') and Mrs Agnes McLehose ('Clarinda'). The elderly Alexander, suffering emotional and physical deprivations after the death of his wife, outlined to Burns his fleshly torments and his design to marry a young beauty who had been second maid of honour to his deceased wife. Burns, in jest or earnest, applauded the suggestion, while 'Clarinda' reacted with horror and puritanical outrage. ('In the name of wonder, how could you spend ten hours with such a— as Mr. Pattison? What a despicable character!') Alexander ultimately abandoned the alluring proposal.

Lt.-Col. Alexander Hope Pattison, by Thomas Duncan. (Glasgow Art Gallery and Museum)

Notes

I have resisted the temptation to justify every fact and statement with a reference note, which in the first draft of the book numbered about 800! The Bibliography contains a full list of all primary and secondary sources; the Notes which follow make reference to short titles only.

Notes to Chapter 1

1. A previous biographer, Miller, questioned Pattison's date of birth because of the latter's own carelessly worded remarks in his *A Refutation of Certain Calumnies*, but the parish entry is unequivocal. His baptism was not recorded.
2. *Glasgow Mercury*, 12 January 1790.
3. Lyle, *Ancient Ballads and Songs*, pp. 228–29.
4. Anonymous, 'The Park District', p. 439.
5. The house was demolished in 1900 to provide space for the International Exhibition of 1901. The grounds are now part of a public park. In Glasgow today the original site of the house seems centrally located, but at the time when the Pattisons were in residence Kelvingrove was more than one mile from the closest part of the city. Nowadays the approach is by Sauchiehall Street which at that time was a narrow, unpaved country lane, full of deep ruts. So bad was it that it could not be used for carriages, hence the route that the Pattisons had to travel was more circuitous but led eventually to the handsome gate and lodge of the estate.
6. The Glasgow Grammar School was an ancient institution, to which reference is made as early as 1461. In 1834 it was renamed the High School. My description of school life is based on Cleland's *The History of the High School of Glasgow*,

pp. 10–13. An anonymous handwritten record provides an outline of the subjects studied at the school: 'Glasgow Grammar School: Course of Study and Principal Class Books'; although dated some twenty years after Pattison's time, it can be considered a fair indication of the courses he would have studied, as the course content in those days changed little from decade to decade. Particular details about Pattison's performance at school are found in his surviving class records.

7. Henry Grey Graham provides a fascinating description of the Candlemas proceedings in his *Social Life of Scotland*, pp. 431–32. He has portrayed the masters sitting, the stern air of authority gone, the instruments of punishment concealed, with a subdued air of expectancy on their countenances. Until 1786 a distasteful and invidious extension of the proceedings had regularly taken place. When the sum given was less than five shillings, no notice was taken, but when it amounted to that sum, the rector said '*Vivat*' (let him live) and all the boys gave one ruff (a round of applause with the feet). Ten shillings merited '*Floreat*' (let him flourish) and two ruffs; fifteen shillings '*Floreat bis*' (let him twice flourish) and three ruffs; twenty shillings '*Floreat ter*' (let him thrice flourish) and four ruffs; and a guinea and upwards '*Gloriat*' (let him be glorious) and six ruffs. When this had all been completed, the rector stood up and in an audible voice declared the '*Victor*', by stating the name of the boy who had given the largest sum. Graham concludes: 'The ordeal was as undignified for the master as it was injurious to the scholars—the crest-fallen, bitter humiliation of the poor lads, the contemptuous purse pride of the rich pupils.'

8. There is a commemorative plaque to David Allison on the wall of Glasgow Cathedral; he died on 9 December 1808 at the age of fifty-eight.

9. It is unlikely that Pattison completed his years of schooling unscathed. On a page of excerpts from a Grammar School Album there appears this ominous minute from the conveners of the committee to the teachers: 'That all the masters, and particularly when they take charge of the young class, endeavour to bring the boys on as regularly as possible, and to bestow some extraordinary pains upon the boys at the lower end of the class in order to bring them up, as far as can be done, with the others.'

10. The abbreviation 'f.n. 5tus' stands for *filius natu quintus* (fifth son)—of John, Merchant, Glasgow, in the county of Lanark. Pattison's name and a biographical sketch appear in Addison's *The Matriculation Albums*, p. 227. Granville Sharp Pattison is entry no. 7263 but is incorrectly listed as 'Georgius Watson' who is entry no. 7262.

11. A great deal of my information about the teaching of medicine at this period derives from *Evidence, oral and documentary*, vol. 2. One of the members of the royal commission which made the report was Lord Justice-Clerk David Boyle, the presiding judge at Pattison's grave-robbing trial in 1814.

12. The evidence for Pattison's list of classes is derived from a variety of sources in the Archives of the University of Glasgow. He matriculated in Professor John Young's Greek class; this was a second-year course of study. The only way to enter directly into second year was by passing an oral public examination in Latin. Since it seems improbable that Pattison would have passed such an examination, it is likely that he had to take a year of Latin in 1806/07 and that therefore he matriculated as a second-year student in 1807. His attendance at all other classes is listed in the Register of Medical Students: 1808/09, p. 58, student no. 174; 1809/10, p. 73, student no. 231; 1810/11, p. 89, student no. 202; 1811/12, p. 106, student no. 259. Independent evidence for his study of anatomy is found in Professor James Jeffray's Classbook ('November 2, 1808, student no. 45, Mr. Granville Pattison, £3.3.0'), and of practice of medicine and materia medica in The Book of Attested Students, pp. 74–76 (14 November 1809) and p. 84 (14 November 1811).

13. John Burns was appointed the first Professor of Surgery at the University in 1815.

14. This and the following quotations come from Lyle's 'University Reminiscences', transcribed and to be published by Dr L. R. C. Agnew, University of California, Los Angeles, California. Lyle was a surgeon and poet. His reminiscences relate to the period 1812–15 while he was a medical student at the University of Glasgow. I am very grateful to Dr Agnew for kindly allowing me to see and to quote from this work. (His comment about Freer is quoted in Agnew's 'Scottish Medical Education'.)

15. 'Absorbents' was the term used for the lymphatic system including the lacteals (intestinal lymphatics that take up chyle).

16. Hunter, A Short History, pp. 53–54.

17. 'Smith', Northern Sketches, p. 95. The paper of the book is water-marked 1807/08; the probable publication date is c. 1811 (according to the records of the Mitchell Library, Glasgow). The true author is thought to have been either J. Gibson Lockhart or John Finlay. David Murray (Memories of the Old College of Glasgow) considered the latter the more likely, and identified Pattison with the character 'Beau Fribble'. The book is now very rare; because of its notoriety it sold out quickly.

18. Armour to Mackenzie, 28 June 1819.

19. A study of advertisements in the Glasgow Herald reveals that the College Street Medical School justified its name only in 1809 (the year that Pattison joined Allan Burns). John and Allan, who had been lecturing in Virginia Street, moved to College Street in October that year. The location of the School was never specified in correspondence and advertisements. But Dr William Mackenzie, founder of the Glasgow Eye Hospital, took over Pattison's rooms at the School in 1819. In the Archives of the Royal College of Physicians and Surgeons of Glasgow are several letters to Mackenzie, c. 1819, all addressed to 10 College Street. The location of 10 College Street was obtained from a map of

13. The Lord-Advocate, as head of the Scottish system of public prosecution of crimes, appoints about six younger advocates as Advocates-Depute; they draft and sign indictments and appear for the prosecution at trials. In important trials, the Lord-Advocate leads for the crown, and the A.D. assists him; this was the case in Pattison's trial.

14. The indictment and other handwritten documents contain certain words and phrases written in larger lettering; these I have italicized.

15. Older than *The Times* of London, the *Glasgow Herald* has been in continuous publication for over 200 years. In 1814 it was published twice weekly, on Mondays and Fridays. Consisting of one double sheet of news and advertisements, it cost 6½d. per issue. Much of the news was lurid. All public executions were described in gruesome and heart-rending detail. Many of the advertisements were those of medical practitioners advertising for medical students, since private tuition was common in those days.

16. Mackenzie, *Old Reminiscences*, vol. 2, p. 486.

17. Ibid.

18. Hume, *Commentaries*, 2nd ed., 1819: vol. 1, p. 84 (footnote); vol. 2, p. 297. 4th (last) ed., 1844: vol. 1, p. 85 (footnote); vol. 2, p. 304. In both of these editions, the date of trial is incorrectly listed as 16 June 1814 in vol. 1, but is correct in vol. 2.

19. Alexander Maconachie was the son of Lord Meadowbank, one of the presiding judges. Home Drummond later became a member of parliament but never a judge. Francis Jeffrey, a friend of Sir Walter Scott, was famous for his writings in the *Edinburgh Review*; he later became a judge, Lord Jeffrey. Some four years after the trial, he wrote a letter strongly supporting Pattison's application for the chair of anatomy at the University of Pennsylvania. John Clerk, the senior defence counsel, was Pattison's second cousin once removed, through Mary Clerk of Penicuik; he had had a brief term as Solicitor-General in 1806, and was later elevated to the bench as Lord Eldin. Clerk was counsel for Pattison, while Cockburn acted for the other three accused. Many of the personalities involved in the trial were again in court at the trial in 1828 of William Burke and his female associate Helen McDougal for murder and delivery of bodies to the anatomists. William Hare, the other half of the notorious Burke-and-Hare team, had turned 'King's evidence' and was immune from prosecution. The presiding judge was Lord Justice-Clerk David Boyle, assisted by Lords Meadowbank the second (Alexander Maconachie) and Pitmilly. Henry Cockburn, a counsel for McDougal, was successful in having her acquitted. Burke was found guilty and was hanged.

20. *Scots Magazine*, vol. 76 (July 1814), p. 553.

21. Minute Book of the High Court; The Book of Adjournal.

22. *Glasgow Chronicle*, 9 June 1814; *Glasgow Courier*, 9 June 1814; *Glasgow Herald*, 10 June 1814. The last of these provides the most detailed account of the trial.

8. *The Emmet*, vol. 2, pp. 29–30. The students gave as the publisher of the full poem 'John Smith, 25 Gallowgate'. John Smith, a well-known Glasgow bookseller, has never had premises in the Gallowgate. The firm's Manager of the Antiquarian Department, Mr Cooper Hay, considers the poem to be fictitious and that it was never published in full.

9. Hood, *Choice Works*, pp. 191–93. 'Mary-bone' refers to Marylebone, a district of London; 'Dr. Vyse' may be an anglicized form of Dr Weiss (a Dr John Weiss practised as an instrument maker in London during the 1820s and was well known in London medical circles); 'Guy's' refers to Guy's Hospital, the London teaching hospital; 'Doctor Bell' to Dr (later Sir) Charles Bell, the Scottish physiologist, anatomist, and surgeon; 'Doctor Carpue' to Dr Joseph C. Carpue, the English anatomist and surgeon; 'Pickford's' to the British removal company; and 'Sir Astley' to Sir Astley Cooper, the English anatomist and surgeon. 'Mary's Ghost' must have been written after Sir Astley Cooper's title was conferred in 1821 but before Sir Charles Bell received his knighthood in 1831, because of the styles of address used in the poem.

10. The handwritten notes of two of the presiding judges, which run to some fifty pages, and the precognitions (pre-trial statements) of some of the witnesses (248 pages) make possible the reconstruction of the crime and trial. The evidence is presented in the form of submissions and examinations of witnesses, resulting in a fragmentary and scattered array of data. The chronology and chain of events are often obscure and occasionally ambiguous or contradictory.

11. The location of Mrs McAllaster's lair was determined from registers of lair transfers. She was buried in the lair of her husband Walter McAllaster, but no lair number was given (*Parochial Registers*, OPR 644/61). On 16 July 1826 Walter McAllaster sold his lair to his brother-in-law Dugald McGregor, but again no lair number was given (*Ramshorn 1817–72, p.* 12). On 22 April 1835 Jean Scott, widow of Dugald McGregor, sold the lair to Robert Corbet, bricklayer, but once again no lair number was specified (*Ramshorn 1818–46,* p. 106). Finally, the number of Robert Corbet's grave was determined from later entries referring to his deceased children and relatives (*Ramshorn 1855–72,* 30 September 1857, 4 March 1861, 20 April 1862, 5 May 1862, 23 July 1867). The description of the lair was '2/46 EMW St. David's' (second lair out from the wall, lair no. 46, east middle wall, St David's, i.e., Ramshorn churchyard). A plan of *c.* 1840 showed the location of the lair. On visiting the churchyard, I located lair 46, but the horizontal gravestone was completely obliterated by several inches of soil. Helped by three young boys and a friendly bystander, I scraped away the soil, exposing the name 'Robert Corbet', visible for the first time in about a century. This is the location of Mrs McAllaster's burial and grave-robbery.

12. Mackenzie, *Old Reminiscences*, vol. 2, p. 485.

c. 1827. I am very grateful to Dr A. M. Wright Thomson for his help in pinpointing the location of the School.

20. Inkle means tape.

21. Burns, *Surgical Anatomy*, p. xxiv.

22. 'Burns's ligament' is the falciform margin of the saphenous opening of the thigh, and 'space of Burns' is the jugular fossa in the temporal bone of the skull. The eponym 'Burns's amaurosis' (post-marital amblyopia) is named after Allan's brother John.

23. There were twenty-one shillings in a guinea, twenty in a pound.

24. Because it is a tradition in British medical circles to address a surgeon as 'Mr', Pattison throughout his life was sometimes referred to as 'Mr' (in his capacity as a surgeon) and sometimes as 'Dr' (in his capacity as an anatomist and physician). The qualifications of surgeons, *Chirurgiae Baccalaureus* and *Chirurgiae Magister*, were not doctorates; originally physicians alone were 'doctors'—a title conferred only by a university, a ruling monarch, or the Archbishop of Canterbury. British surgeons, by being trained as apprentices of barber-surgeons, were addressed as 'Mr', and they prefer to retain this title even today. Some acknowledge that the practice is a harmless form of inverted snobbery.

Notes to Chapter 2

1. Hunter, *A Short History*, p. 22.

2. Mackenzie, *Old Reminiscences*, vol. 2, pp. 473–75. Mackenzie's *Reminiscences*, adorned with colourful embellishments, are often not strictly correct, but convey the spirit of the episodes he portrays. Their accuracy has been challenged by Ted Ramsey, *Glasgow Herald*, 20 July 1983. (Mackenzie was only a boy at the time of Pattison's trial.)

3. *The Oxford English Dictionary* gives 1776 as the first occurrence of the word 'resurrectionist'. Some thirty-four years earlier the *Scots Magazine* (vol. 4 [March 1742], p. 141) made reference to 'Resurrection Hall, as if built by the gains of that unlawful traffick'.

4. *Report from the Select Committee on Anatomy*, p. 69.

5. Ibid., p. 85.

6. Mackenzie, *Old Reminiscences*, vol.2, pp. 475–78.

7. In England, the removal of a naked body was a misdemeanour, while taking the grave clothes was a felony (theft). Grave-robbers in England were careful to take only the body, because the sentence for a felony was very severe. In Scots Law there was and is no distinction between felonies and misdemeanours, and the two words are not even used. It therefore did not make any difference in Scotland to the charge of violating sepulchres (*crimen violati sepulchri*) whether the grave clothes were taken or not. But even in Scotland clothes were usually replaced in the grave to avoid possible later identification and incrimination.

23. Ibid. *Scots Magazine*, vol. 76 (July 1814), p. 553. *Glasgow Courier*, 11 June 1814. *Edinburgh Evening Courant*, 9 June 1814.
24. Mackenzie, *Old Reminiscences*, vol. 2, p. 488. See Note 2.
25. Ibid.
26. Lord Justice-Clerk David Boyle's words were given in *oratio obliqua* but I have taken the liberty of transposing them into *oratio recta*.
27. *Glasgow Herald*, 10 June 1814. *Glagow Courier*, 11 June 1814.
28. Ibid.
29. Hume, *Commentaries*, 2nd ed., 1819, vol. 1, p. 84; 4th ed., 1844, vol. 1, p. 85.
30. I am very grateful to Professor David M. Walker, Q.C., Regius Professor of Law, University of Glasgow, for providing me with a 'Note on Punishment for Violation of Supulchres'.
31. *Report from the Select Committee on Anatomy*, p. 70.

Notes to Chapter 3

1. Grahame and Mitchell, Letter Book.
2. In 1866 the Glasgow Medical Society joined with a younger society, the Medico-Chirurgical Society of Glasgow, and adopted its name. In 1919 it assumed its present title, the Royal Medico-Chirurgical Society of Glasgow.
3. Graham to Mackenzie, 26 November 1814.
4. *Report of the Committee of the Town's Hospital*, p. 17.
5. Buchanan, *History*, pp. 60–66.
6. It is possible to reconstruct the chain of events from four handwritten documents: Proceedings of the Committee, 2 January 1817; Glasgow Royal Infirmary Records, December 1816 and January 1817; Pattison, Memorial of Exculpation; George Watson, notes. Additional information, comments, and side issues are given in two contemporary letters: Armour to Mackenzie, 21 January 1817; Rainy to Mackenzie, 12 February 1817.
7. Anderson's Institution (or Andersonian Institution) was later known as Anderson's University or the Andersonian. It subsequently became incorporated into the Glasgow and West of Scotland Technical College, which in turn gave rise to the Royal Technical College and finally the University of Strathclyde.
8. Muir, *John Anderson*, p. 145.
9. Minutes, Anderson's Institution, 18 and 24 November 1818, pp. 135–38.
10. The original documents of the Ure divorce trial before the Commissary Court, from which I have taken many extracts, are in the Scottish Record Office, Edinburgh. (Part of these records were printed by Nathaniel Chapman in Philadelphia in order to defame Pattison.)
11. 'Smith', *Northern Sketches*, p. 17.
12. Chapman, *Correspondence*, pp. 21–22.
13. Rainy to Mackenzie, 9 March 1819.

14. Ibid.
15. Pattison, *A Final Reply*, p. 6.
16. Minutes, Anderson's Institution, 13 May 1819.
17. Ibid.
18. Armour to Mackenzie, 28 June 1819.
19. The *Courier* was one of the famous early Black Bull packet ships, each weighing 500 tons, which, independent of weather or freight, sailed punctually on the first day of each month. Their average westbound time was forty days, and eastbound twenty-three days.

Notes to Chapter 4

1. William Potts Dewees had been a candidate for the chair of midwifery at the University of Pennsylvania in 1811, but to his severe disappointment Thomas Chalkley James was appointed instead. Dewees served as adjunct professor until his succession to the chair in 1834. See Corner, *Two Centuries*, p.86.
2. Chapman, *Correspondence*, pp. 12–13.
3. Pattison, *A Refutation*, p. 12.
4. Both Chapman and Pattison wrote accounts of the quarrel. In these, many stated facts and opinions are contradictory and unverifiable. I have relied more on Pattison's narrative because subsequent events have confirmed at least some of his assertions. See *A Refutation* and *A Final Reply*.
5. Pattison, *A Refutation*, p.16.
6. 'Commonwealth vs. Granville Sharp Pattison, 26 October 1820, for posting. Ignoramus, county pay costs, defendant discharged.' Mayor's Court Docket. (*Ignoramus*, we ignore, was formerly written by a grand jury on the back of a rejected indictment.)
7. Pattison, *A Refutation*, p.22.
8. Chapman, *Correspondence*, p.33.
9. Chapman in his *Correspondence*, p.34, acknowledged that he had challenged Dewees, no reason being given, and stated that this was in 1805 or 1806, when Dewees was childless. Pattison then showed in *A Final Reply* that the challenge was issued in 1811, when Dewees indeed had two or three children.
10. Pattison, *A Refutation*, p.21.
11. Dunglison, 'Autobiographical Ana', p.84.
12. Richman, *The Brightest Ornament*, p.164.
13. 'X' is incorrectly referred to as 'Z' in some subsequent articles.
14. Chapman, *Correspondence*, p.28.
15. Quoted in Pattison, *An Answer*, postscript, p.32.
16. Hunter, *A Short History*, p.86.

17. The Burns Club was independent of the St Andrew's Society of Philadelphia. The former was a short-lived social organization motivated by the literary interests of the members, whereas the latter, founded in the eighteenth century, was stable and prestigious. Gibson was a member of the St Andrew's Society and Chapman was its physician; it is easy to see why Pattison, although an educated Scotsman, never became a member. I am very grateful to Dr F. J. Dallett, University of Pennsylvania Archivist, for this information.

18. Porter to Miner, 25 October 1825.

19. Pattison, *A Refutation*, p.53.

20. The original thesis is lodged in the Van Pelt Library, University of Pennsylvania, call no. 378.748/POM2. Parts of pages 16 and 20 are crossed out; pages 17–19 were missing until recently. An explanation was added to the thesis on the back of page 16: 'The pages here removed contained sundry observations on the subject of nosology, which the medical faculty deemed to be improper and not to be sanctioned by them. With this impression, Mr. Whilldin was required on the 24th instant to eradicate the parts disapproved of—which was accordingly done by himself, in presence of the faculty.'

21. The portrait by Chester Harding, painted in Scotland in 1826, was shown at the Exhibition of Portraits, Glasgow, 1868, and reproduced in the catalogue. I do not know its present location. An engraving 'painted by C. Harding, engraved and printed by J. Sartain' can be seen in the Library of the Royal College of Physicians, London. I am grateful to Mrs Leah Lipton, Framingham State College, Massachusetts, for helpful discussions.

22. Pattison, *A Refutation*, p.21.

23. Pattison's interest in electricity may have been sparked by Ure's and Jeffray's experiments on the executed Clydesdale's body, which he probably witnessed in 1818 while living in Glasgow. See Pattison's letter to Gilmour.

24. The present custodian of the museum is Mr Ronald Wade, Department of Anatomy, University of Maryland, to whom I am very grateful for help and advice.

25. Shryock, *Modern Medicine*, p.246.

26. Callcott, *A History*, p.41.

27. *Federal Gazette and Baltimore Daily Advertiser*, 10 February 1824.

28. Gross, *Autobiography*, vol. 1, p.166; vol. 2, p.257.

29. University of Maryland, 'Minutes of the Executive Committee', 7 March 1837, pp.53–54.

30. Callcott, *A History*, p.124.

31. Cordell, *University of Maryland*, vol. 1, p.163.

32. Pattison, *A Lecture delivered in Jefferson Medical College*, p.16.

33. Cordell, *Historical Sketch*, p.53.

34. Nicholas to Cadwalader, 9 April 1823.

35. Scott to Cadwalader, 15 April 1823.

36. Callcott, *A History*, p.50.

37. By her second marriage, Mrs Patterson became Marchioness of Wellesley and Vicereine of Ireland, and the sister-in-law of the Duke of Wellington.
38. Quoted in Callcott, *A History*, p.53.
39. Pattison to Meredith, 'Off the Capes of Delaware'.

Notes to Chapter 5

1. Carlisle, Sir Anthony, quoted in Newman, *Medical Education*, p.184.
2. Quoted in Newman, *Medical Education*, p.43.
3. Ibid.
4. A royal charter was finally granted on 28 November 1836. From this point on the institution came to be known as 'University College, London', and formed a constituent part of the new and larger 'University of London'. Until 1836 London was the only capital city in Europe without a university. (Two books provide an excellent account of the early years: Bellot's *University College, London* is long, detailed, and factual, while Harte and North's *The World of University College* is lively, with a wealth of interesting illustrations.)
5. *Statement by the Council*, p.23.
6. *London University Calendar*, pp.114–17, 150–52.
7. Pattison to council, 18 December 1827.
8. *London University Calendar*, p.117.
9. *Lancet*, 1826–27, xii, 18.
10. Pattison, *Statement of the Facts*, p.4.
11. Dr George Birkbeck, after whom Birkbeck College, University of London, is named, was the only member of the university council with medical training. He was a consistent supporter of Pattison and a sore trial to Leonard Horner, the warden.
12. Pattison, *Statement of the Facts*, pp.4–10 contain a description of the incident and its consequences.
13. Mr (later Sir) Charles Bell was an outstanding anatomist, surgeon, and physiologist of established fame. He is remembered for the eponym 'Bell's palsy' (facial nerve paralysis). John Conolly had been a student of Pattison's in 1818 in Glasgow and was at all times a loyal supporter, both in Glasgow and London. He became a well known authority on insanity and was a pioneer in developing humane methods of treatment in mental asylums, including the abolition of mechanical restraint of the insane.
14. Bennett to Horner, 22 August 1829.
15. Pattison to council, 11 September 1829.
16. Pattison to council, 5 December 1829.
17. Pattison, *Statement of the Facts*, p.11.
18. Davis, one of Pattison's consistent supporters, attended the Duchess of Kent at the birth of Queen Victoria in 1819.
19. 'Report of Professors Conolly et al.', 9 December 1829.

NOTES TO pp. 154–173

20. Bennett to council, 16 January 1830.
21. University College Hospital, originally 'North London Hospital', opened in 1834.
22. 'Report of Professors Conolly *et al*.', 10 February 1830.
23. Pattison to council, 14 February and *c*. 16 February 1830.
24. *Lancet*, 1829–30, i, 897–98. In a later issue, the writer was identified as Nathaniel Eisdell: *Lancet* 1829–30, ii, 623–24.
25. A number of dissident medical students were particularly persistent in their attacks on Pattison. Their names, in decreasing significance, were Nathaniel Eisdell, Alexander Thomson (M.B., Cantab.), J. H. Peart, Charles Robert Bree, J. Merriman, Edward Meryon, Henry Thomas, John Rayner, John Bartlett, and Thomas Evans Brinsden.
26. Pattison, *Statement of the Facts*, p.14.
27. Ibid., p.15.
28. The four students (anonymous in the report) were later identified as Nathaniel Eisdell, William Jorden, Charles Nelson, and Henry Plank. See Thomson, *To Lord Brougham*, p.23.
29. Pattison to council, undated, July 1830.
30. Pattison to council, 19 June 1830.
31. Professor Thomson had no control or influence over his son Alexander and relations between father and son were strained. Both later united in their opposition to Pattison.
32. Pattison, *Lancet*, 1831–32, i, 209–15.
33. 'Recent Improvements'.
34. Pattison to council, 5 September 1830.
35. *Lancet*, 1830–31, i, 286–87.
36. Pattison to council, 22 November 1828.
37. Pattison, *Statement of the Facts*, p.22.
38. Fearon to council, 28 February 1831.
39. Pattison to council, 7 March 1831.
40. Pattison to council, 5 March 1831.
41. Pattison, *Statement of the Facts*, p.26.
42. Council 'Minutes', vol.2, 30 April 1831, p.271. The students were Brinsden, Bartlett, and Rayner.
43. Pattison, *Statement of the Facts*, p.28.
44. Pattison to council, 7 May 1831. Council 'Minutes', vol.2, 16 May 1831, p.276.
45. 'Memorial to the Council', signed by six professors, *c*. May 1831.
46. Council 'Minutes', vol.2, 23 July 1831, p.308.
47. Pattison to council, 16 December 1831.
48. In the midst of much turmoil, he wrote to the clerk of the council: 'Tuesday is Michaelmas Day and children make a great account of ceremonies on these days, so pray tell our cook to have a roast goose and apple pie for them.' Horner to Coates, 26 September 1829.

49. Cockburn, *Life of Lord Jeffrey*, vol. 1, p.210.
50. *Sun*, 22 April 1830.
51. Horner to council, 7 March 1831.
52. Pattison to council, 19 March 1831.
53. Bell, *Letters*, p.317.
54. Horner duly recovered and later played an important role in factory legislation. It is interesting that Andrew Ure and Leonard Horner, Pattison's major antagonists in Glasgow and London respectively, were themselves pitted against one another in the formulation of factory legislation. Ure favoured child labour and published lyrical descriptions of the children's happy lot; Horner, as a factory inspector following the passage of The Factory Act of 1833, recognized and drew attention to the true state of affairs in his reports on the unbelievable cruelty to and abuse of children. (See Pike, *Human Documents*. pp.154–62, 214, 223.) At the time that Pattison was elected to the chair of anatomy at London, Ure was an unsuccessful candidate for the chair of chemistry, losing to Edward Turner.
55. de Morgan, *Memoir*, pp.34–37.
56. *Lancet*, 1830–31, ii, 695.
57. *Lancet*, 1830–31, ii, 793–95.
58. Turner and Thomson, *Lancet*, 1830–31, ii, 744–50.
59. *Statement by nine professors*, p.20.
60. Wakley, *Lancet*, 1830–31, ii, 787.
61. de Morgan, *Memoir*, pp.34–37.
62. Dunglison, 'Autobiographical Ana', p.113.
63. Featherstonhaugh to Murchison, 28 October 1831.
64. Pattison, *Lancet*, 1831–32, i, 215.
65. *Lancet*, 1831–32, i, 812.

Notes to Chapter 6

1. Sims, *The Story of My Life*, p.137.
2. Osler, *An Alabama Student*, pp.233–34. Stillé misquoted Byron's lines on the skull: 'It is a temple where a god might dwell—a house not made with hands, but whose architect is the architect of the universe.'
3. *Announcement*, session MDCCCXXXVI-VII, pp.5–6.
4. *Register and Library of Medical and Chirurgical Science* 1, no. 45 (1834): 386–407.
5. Gross, *Autobiography*, vol. 2, pp.256–60.
6. Sims, *The Story of My Life*, pp.131–32.
7. Dunglison, 'Autobiographical Ana', pp.84–86. Dunglison wrote his 'Autobiographical Ana' (i.e., a collection of anecdotes, reminiscences, etc.) to preserve recollections and records thought to be of interest to his family. To the original eight holograph volumes he added supplementary notes from time to time.

8. Ibid., pp.87–88.
9. William Robert(s) Jones attended the Jefferson Medical College in the period 1830–36. He had previously attended Pattison's anatomy classes at the University of London in 1829–31 and was one of the forty-one students who had signed a memorial in support of Pattison on 5 March 1831. Jones moved to Philadelphia in 1832 as Pattison's apprentice, and graduated at the college in 1834. In spite of the unexplained rift, Pattison wrote to the council of the University of London the following year requesting free tickets for Jones to enable him to practise as a surgeon in England.
10. Dunglison, 'Autobiographical Ana', pp.87–88.
11. The company started as an unincorporated private partnership managed by a board of directors.
12. Pattison to Jaudon, 30 April 1837. The years 1825/26 saw an economic depression in both Britain and the United States, but milder than that of 1837.
13. Wisconsin Tribune, 15 December 1848.
14. Bauer, Doctors Made in America, p.59. 'Depurated' means purified; by implication, the faculty was dissolved.
15. John Revere was the youngest son of Paul Revere, the American patriot of Revolutionary fame.
16. Dunglison, 'Autobiographical Ana', p.94.
17. Ibid. Dunglison's opinion of Pattison and Revere was further tarnished after they moved to New York University. In the first announcement of its new medical faculty, they sanctioned the totally unfounded assertion that the greater part of the museum of Jefferson Medical College belonged to them and that they were bringing it with them. The true state of affairs was that, when their share was allotted to them, it was so insignificant that it was sent to a local Philadelphia apothecary for sale. Pattison in turn had reason to doubt Dunglison's discretion when, at a valedictory lecture in 1841, the latter betrayed a confidence when he announced that the Chancellor of New York University, the Rev. Dr Mathews, in 1839 had offered him a professorship, which he had declined. See Pattison to Mathews, 27 February 1841 and Dunglison, 'Autobiographical Ana', p.91.
18. During Pattison's time, the university was referred to as the University of New York or, occasionally, the University of the City of New York.
19. The clinic was always referred to as the clinique in contemporary accounts.
20. Gross, Autobiography, vol. 2, p.258; Satchwell, Annual Address, p.6.
21. Dunglison, 'Autobiographical Ana', p.98.
22. Ibid, p.97.
23. Satchwell, Annual Address, pp.6–7.
24. Gross, Autobiography, vol. 2, p.257; idem., 'A Century of American Medicine'.
25. The content of Pattison's lectures can be seen in the notes of W. G. Henderson and two unidentified students. Henderson, not one of his better students, kept poor notes abounding in spelling mistakes ('musels', 'sirgury', 'coledge', etc.).

26. *New York Herald*, 10 December 1841.
27. Gross, *Autobiography*, vol. 2, p.259.
28. The twenty-four-year-old Prince Ferdinand de Joinville, third son of King Louis Philippe of France, carried Napoleon Bonaparte's remains from St Helena to Paris.
29. The *New York Herald*, 16 August, 15 November, 12 October 1841, reported these incidents at the clinic.
30. A few anecdotal accounts of major surgery written from the patients' viewpoint have been written. See Graham, *Surgeons All*, pp.314–15; Brown, *Horae Subsecivae*, vol. 2, pp.376–79; Bishop, *Early History of Surgery*, pp.133–34.
31. *New York Herald*, 26 July 1841. Pattison was identified as the surgeon who wielded the 'long, glittering knife' in the *New York Herald*, 13 September 1841.
32. *New York Herald*, 13 September 1841.
33. Pattison, 'Removal of a Carcinomatous Tumor'.
34. Chamberlain, *Universities and Their Sons*, p.98.
35. *New York Herald*, 6 January 1842.
36. Ibid., 13 January 1842.
37. Ibid.
38. Ibid., 18 January 1842.
39. *New York Lancet*, 1 (1842): 218.
40. *La Lancette Française*, 19 February 1842, p.103.
41. *New York Lancet*, 2 (1842): 283.
42. Ibid., 1 (1842): 251, 284.
43. Ibid., 2 (1842): 186.
44. Ibid., 2 (1842): 187.
45. Pattison to Bradish, 6 June 1850.
46. Dunglison, 'Autobiographical Ana', pp.97–98. The cause of death described in Clymer's letter is not entirely clear; by the 'common duct' he probably meant the bile duct (also known as the common bile duct or ductus communis choledochus); it is less likely that he meant the common hepatic duct. In some obituaries, Pattison's death was indeed attributed to obstruction of the ductus communis choledochus but in others to obstruction in the hepatic duct from the presence of biliary calculi. No mention is made of other possible associated conditions such as pancreatitis or carcinoma of the head of the pancreas. From the available evidence, it is probable that the cause of death was rupture of the bile duct secondary to blockage by gallstones.
47. The location of Pattison's lair in the Glasgow Necropolis is beta 3. The date of reinterment was 26 March 1852.
48. The will and a codicil, dated 29 March 1848 and 6 June 1851 respectively, were probated on 22 January 1852 with three of Pattison's nephews, all sons of his brother John, acting as executors. Witnesses included Henry A. Mott and Gunning Bedford.

49. Pattison, *A Lecture delivered in Jefferson Medical College*, p.16. The quotation, which means 'Truth is mighty and will prevail', contains a vulgar Latin form—*prevalebit* for *praevalebit*. The axiom was delivered, perhaps by a Roman, but certainly by Thomas Brooks in *The Crown and Glory of Christianity*, p.407.

Bibliography

Primary Sources

Correspondence and manuscripts for which no location is given are in the Archives of the University of London, D. M. S. Watson Library (Manuscripts and Rare Books Room), University College, Gower Street, London. Most of these are from the 'Pattison Case Papers' (115 documents) or from the relevant College Correspondence (56 documents).

Act for consolidating and amending the Statutes in England relative to Offences against the Person (1828). The Statutes of the United Kingdom of Great Britain and Ireland. 9 George IV, 1828. Cap. XXXI, pp.97–110.

Act for regulating Schools of Anatomy (1832). Public General Statutes. 2 and 3 William IV, 1832. Cap. LXXV, pp.535–39.

Act of Incorporation and Supplementary Acts, with the By-laws of the Medical and Chirurgical Faculty of Maryland, 1848 (Baltimore: Matchett, 1848).

Act of the Scottish Parliament, 1693, c. 27.

Act to amend the Act of the second and third years of William the Fourth, chapter seventy-five, for regulating Schools of Anatomy (1871). 34 and 35 Victoria, 1871. Cap XVI, p.129.

Agreement between G. S. Pattison and J. R. Bennett, 2 August 1830.

American and Commercial Daily Advertiser, Baltimore, 24 August 1820, 25 October 1820, 23 April 1821, 30 September 1826.

American Daily Advertiser, 2 February 1820, 3, 4, 6 and 8 March 1820, 5 April 1820, 4 June 1833.

Anderson's Institution, Minutes of the Meetings of the Managers and Trustees of, 1811–30. Archives of the University of Strathclyde, Glasgow.

Armour, James to William Mackenzie, 21 January 1817; 28 June 1819. Archives of the Royal College of Physicians and Surgeons of Glasgow.

'Articles of Agreement entered into by and between the Managers of the Infirmary in Baltimore and the underwritten Superior General of the Ladies incorporated under the name of Sisters of Charity of St Joseph's, 1823.' Archives, Daughters of Charity, Saint Joseph's Provincial House, Emmitsburg, Maryland.

Atlas, vol. 5, 7 March 1830, p.154.

Bell, Charles to the council of the University of London, 9 September 1829.

Bennet, John (Procurator Fiscal, Glasgow) to Hugh Warrender (Crown Agent), 4 March 1814. Scottish Record Office, Edinburgh, AD (Lord Advocate's Department) 14/14/29.

Bennett, James Richard to the council of the University of London, undated, accepting the appointment of the demonstratorship of anatomy; 27 August 1829; 15 January 1830; 16 January 1830.

————to Leonard Horner, undated, headed 'Tuesday'; 25 March 1828; 22 August 1829, quoted in Horner, *Letter to the Council*, pp.8–9.

Bennett, Mary and Eliza Bennett to the council of the University of London, 14 May 1831.

Book of Adjournal, 6 and 7 June 1814. Scottish Record Office, Edinburgh, JC (Justiciary Court) 4/7.

Boyle, David, Lord Justice-Clerk, Justiciary Note Books of Lord Justice-Clerk David Boyle, 1814, 'Violating Sepulchres of the Dead'. National Library of Scotland, Adv. MS. 36.3.1. vol.1, pp.441–71.

Cadwalader, Thomas, bundle of twelve letters entitled 'Affair with Professor Pattison'. Cadwalader Collection, Thomas Cadwalader Section, Cadwalader Papers 1823, Box 7T, Historical Society of Pennsylvania, Philadelphia.

Church of the Ascension, Fifth Avenue at Tenth Street, New York, Church Records, 13 November 1851.

Cleghorn, Dr Robert, Dr Robert Watt and William Anderson Surgeon, Report of, 17 December 1813. Scottish Record Office, Edinburgh, AD 14/14/29.

Cockburn, Henry, Submission from. Scottish Record Office, Edinburgh, Small Papers, JC 26/368.

Colquhoun, Archibald (Lord Advocate) to Hugh Warrender, 4 February 1814. Scottish Record Office, Edinburgh, AD 14/14/29.

Columbian Observer, 10 April 1823.

Committee Report, undated and unsigned, written on the order of the council of the University of London, 12 June 1830.

Contract of copartnery between Andrew Russel and Granville Sharp Pattison, 4 August 1813. Scottish Record Office, Edinburgh, Small Papers, JC 26/368.

Davis, David D. to the council of the University of London, 6 June 1831.

Deeds, Register of (Durie), vol.320. Scottish Record Office, Edinburgh.

DuBois, Father John to Mrs Mary Patterson, 10 May 1822. Archives, Daughters of Charity, Saint Joseph's Provincial House, Emmitsburg, Maryland.

DuBois, Father John to Mrs Mary Patterson, 9 June 1822. Quoted in A Daughter of Charity, *Mother Rose White*.

————to Granville Sharp Pattison, 7 January 1824. Archives, Daughters of Charity, Saint Joseph's Provincial House, Emmitsburg, Maryland.

Edinburgh Evening Courant, 9 June 1814.

Eisdell, Nathaniel to the council of the University of London, 30 April 1830; 5 May 1830; 15 July 1831.

Faculty of Physicians and Surgeons of Glasgow, Minute Book, 1807–20. Archives of the Royal College of Physicians and Surgeons of Glasgow.

————, receipts and letters. Archives of the Royal College of Physicians and Surgeons of Glasgow.

Fearon, Herbert B. to the council of the University of London, 28 February 1831.

Featherstonhaugh, G. W. to Sir Roderick Impey Murchison, 28 October 1831. American Philosophical Society, Philadelphia.

Federal Gazette and Baltimore Daily Advertiser, 9, 10, 15 and 17 May 1821, 31 October 1821, 15 November 1821, 30 March 1822, 24 October 1823, 10 February 1824, 27 June 1827, 20 July 1830.

First Presbyterian Church, Baltimore, 'Record of Baptisms, Marriages and Deaths, 1767–1879', 17 January 1823.

Glasgow Chronicle, 9 June 1814.

Glasgow Courier, 9 June 1814, 11 June 1814.

Glasgow Directory, The ('Post Office Glasgow Directory').

Glasgow Exhibition, Catalogue of Old, Institute of the Fine Arts, Glasgow (1894).

Glasgow Grammar School Class Book, vol.3, 1800–05. Mitchell Library, Glasgow, High School (of Glasgow) papers, MS 140/16/2.

'Glasgow Grammar School: Course of Study and Principal Class Books', 18 October 1825. Mitchell Library, Glasgow, High School (of Glasgow) papers, MS 140/1/2.

Glasgow Grammar School, Excerpts from 'The Album of the Grammar School'. Mitchell Library, Glasgow, High School (of Glasgow) papers, MS 140/1/7.

Glasgow Herald, 26 October 1807, 20 October 1809, 23 September 1811, 25 October 1811, 5 October 1812, 29 October 1813, 24 December 1813, 25 April 1814, 10 June 1814, 7 October 1814, 29 September 1815, 6 November 1815, 20 November 1818, 27 August 1819, 9 October 1819, 20 July 1983.

Glasgow Medical Society, Medical Essays, 17 January 1815 ff. Archives of the Royal College of Physicians and Surgeons of Glasgow.

————, Minute Book. Archives of the Royal College of Physicians and Surgeons of Glasgow.

Glasgow Mercury, 12 January 1790.

Glasgow Necropolis, Register of Burials, 1833–54.

Glasgow Royal Infirmary, the Twenty-Second Annual Report, for the year 1816; the Twenty-Fourth Annual Report, for the year 1818. Glasgow Royal Infirmary Archives.

Glasgow Royal Infirmary, 'Proceedings of the Committee on Mr. Pattison's
Matter, 2 January 1817'. Archives of the Glasgow Royal Infirmary.
Glasgow Royal Infirmary Records, 1 May 1815; December 1816; January 1817.
University of Glasgow Archives, HB 14/1/3.
Glasgow University, 'Book of Attested Students, 1766–1843'. University of Glasgow
Archives.
————, 'Matriculation Album', University of Glasgow Archives, F263/1807
GUA 26678 (Clerk's Press No.64).
————, 'Register of Medical Students in the University of Glasgow, begun on
the 21st of January, 1804', vol.1. University of Glasgow Archives.
Glasgow University Calendar for the Session 1826–27 (Glasgow: Knull, 1827).
Graham, Robert to William Mackenzie, 26 November 1814. Archives of the Royal
College of Physicians and Surgeons of Glasgow.
Grahame and Mitchell, Letter Book, 1814–15. Glasgow City Archives, T-MJ 56.
Grant, Robert E., Edward Turner and A. Todd Thomson to council, 6 June 1831.
Hare, Robert to Granville Sharp Pattison, 17 November 1820. B.H22, The
American Philosophical Society, Philadelphia.
Henderson, W. G., 'Notes on Professor Pattison's Lectures, 1840–41'. Archives of
the College of Physicians of Philadelphia.
High Court, Minute Book of, 6 and 7 June 1814. Scottish Record Office,
Edinburgh, JC 8/10.
Horner, Leonard to Thomas Coates, 26 September 1829.
————to the council of the University of London, 5, 7, 10, 11 and 23 March
1831; 9 July 1831.
————to Granville Sharp Pattison, 30 April 1831.
Indictment, His Majesty's Advocate agt. Pattison and others, 1814. Scottish Record
Office, Edinburgh, Small Papers, JC 26/368
Iowa Copper Mining Company. (Unless otherwise indicated, all papers relating to
the company—legal documents, court material, deeds, receipts and much
correspondence—are housed in the Detroit Public Library, MSS 68–199,
Burton Historical Collection; Bank of the United States, Pennsylvania, 1836–
41.) John Andrews to Charles S. Hempstead, attorney, 11 March 1840. John
Bracken, undersheriff of Iowa County, Wisconsin: public notices, 24 November
1840 and 9 March 1841, of a mortgage sale of the company's property and
copper smelting furnace (newspaper entries). B. B. Burrows to T. S. Stevens, 13
December 1839. John Cadwalader to the Bank of the United States, 29 April
1841. Francis J. Dunn to John Bracken, 17 April 1844, 7 June 1844, 3 June 1846.
Indenture, 26 January 1837, regarding the purchase by Pattison and John
Andrews of land from Benjamin and Eliza Mills and John D. Ansley. Granville
Sharp Pattison to John Andrews, 31 March 1837 (Quaker Collection, no.231,
Haverford College, Haverford, Pennsylvania), 10 December 1839 (Simon Gratz
Collection, case, 7, box 32, Historical Society of Pennsylvania, Philadelphia), 12

December 1839 (Dreer Collection, Physicians and Surgeons, Historical Society of Pennsylvania, Philadelphia). Granville Sharp Pattison to Samuel Jaudon, 30 April, 22 May and 6 June 1837. Granville Sharp Pattison and John Andrews, letter of attorney to John Bracken, 10 March 1840. Shareholders of the Iowa Copper Mining Company.

Jefferson Medical College, Philadelphia, Annual Announcement of Lectures, etc. by the Trustees and Professors of, for the year MDCCCXXXII; and session MDCCCXXXVI-VII. Office of the Registrar, Thomas Jefferson University, Philadelphia.

Jefferson Medical College of Philadelphia, Catalogue of Graduates of, *1826–1879*. Office of the Registrar, Thomas Jefferson University, Philadelphia.

Jefferson Medical College, Philadelphia, Minutes of the Board of Trustees, 16 August 1826 to 19 February 1840; 19 April 1838 to 26 November 1873. Office of the President, Thomas Jefferson University, Philadelphia.

Jeffray, James, Classbook of Anatomy. University of Glasgow Archives.

Lanark County, Barony Parish Glasgow, Register of Births and Baptisms, vol.3 (1782–98), January 1791. Scottish Record Office, Edinburgh.

Lanark, County of, Glasgow, Parochial Registers, North West and Blackfriars, 1788–1808. New Register House, Edinburgh, OPR 644/60 (Microfilm, Mitchell Library, Glasgow, G929.3, B832123–5), 4 January 1808.

————, Parochial Registers, Ramshorn and Blackfriars, 1808–1819. New Register House, Edinburgh, OPR 644/61. (Microfilm, Mitchell Library, Glasgow, G929.3, B832123–5).

La Lancette Française: Gazette des Hôpitaux Civils et Militaires, 19 February 1842, p.103.

Lentz, Robert T., Archivist, Thomas Jefferson University, Philadelphia, private communication to the author, 20 July 1978.

London University Calendar for the year MDCCCXXXI, The (London: Taylor, 1831).

Lord Lyon King of Arms, Matriculation Records, 16 December 1802, vol.1, folio 588; 22 March 1962, vol.46, p.94.

Lyle, Thomas, 'University Reminiscences', *c.* 1833, manuscript owned by Dr L. R. C. Agnew, University of California, Los Angeles.

Mayor's Court Docket (Criminal Court), September 1819–June 1822, p.287, entry no. 150. Department of Records, Archives of the City and County of Philadelphia.

McAllaster, Walter, Merchant in Glasgow, Memorial for, 24 June 1814. Scottish Record Office, Edinburgh AD 14/14/29.

McLean, John, Declaration of, 14 December 1813. Scottish Record Office, Edinburgh, Small Papers, JC 26/368.

Memorial to the council of the University of London, signed by seventeen students, 15 May 1830.

Memorial to the council of the University of London, signed by sixty students, 15 February 1831.

Memorial to the council of the University of London, signed by forty-one students, undated, received by council on 5 March 1831.

Memorial to the council of the University of London, signed by eighty-nine students, 16 March 1831. (This memorial is identical to that signed earlier by sixty students.)

Memorial to the council of the University of London, signed by six professors, undated, probably May 1831. (One of the professors was Augustus de Morgan.)

Memorial to Granville Sharp Pattison, undated, c. January 1830, signed by seventy-three students.

Monro, Robert, Declaration of, 14 December 1813. Scottish Record Office, Edinburgh, Small Papers, JC 26/368.

Morning Chronicle, 5 July 1830, 22 August 1831, 5 September 1831, 14 September 1831.

New York City Directory, Doggett's, 1842–52.

New York Directory for the 66th and 67th Years of Independence, Longworth's, 1841–43.

New York Herald, 22, 26, 31 July 1841; 9, 16, 23, 30 August 1841; 6, 13, 20, 27 September 1841; 4, 12, 18, 23, 25, 26, 27, 28, 29 October 1841; 1, 3, 10, 15, 29 November 1841; 6, 10, 12, 19 December 1841; 6, 8, 9, 10, 13, 17, 18 January 1842; 28 February 1842; 18 March 1842; 6, 12, 21, 26 April 1842; 13 November 1851.

New York Post, 1 June 1833.

Nicholas, Charles J. to Thomas Cadwalader, 9 April 1823. Cadwalader Collection, Thomas Cadwalader Section, Cadwalader Papers 1823, Box 14T, Historical Society of Pennsylvania, Philadelphia.

Niles Weekly Register, Baltimore, vol.9, 15 September 1815, pp.34–35.

'Notes of a Viva Voce Examination before a Committee of the Council, 12 July 1830, of four medical students.'

'Notes on the Trial of Granville Sharp Pattison and Others' (the writer may have been one of the presiding judges). Edinburgh University Library, Dk.1.27, presented by Dr J. Menzies Campbell, 1960.

Paisley Abbey, Renfrewshire, Monumental Inscriptions, 29 August 1786 and 9 December 1786.

Pattison, Alexander Hope, handwritten notes relating to his great grandfather, John Pattison of Kelvingrove and family. In the author's possession.

Pattison, Granville Sharp, Declaration of, 14 December 1813. Scottish Record Office, Edinburgh, Small Papers, JC 26/368.

———, Second Declaration of, 21 January 1814. Scottish Record Office, Edinburgh, Small Papers, JC 26/368.

———, 'Last will and Testament'. New York County Surrogate's Court, 31 Chambers Street, New York: Liber 103, pp.215–21 (probated on 22 January 1852).

———, Memorial of Exculpation (very rough first draft) c. 2 January 1817, 20pp. Archives of the Glasgow Royal Infirmary.

Pattison, Granville Sharp, *Testimonials transmitted to the Council of the University of London* (London, 1830).

————to J. R. Bennett, 1 November 1828.

————to Luther Bradish, 6 June 1850. Miscellaneous MSS, New York Historical Society.

————to Messrs Carey and Lea, 7 July 1822. Historical Society of Pennsylvania, Philadelphia.

————to Thomas Coates, Secretary to the Council of the University of London, 8 December 1835.

————to the Council of the University of London, 18 December 1827; 22 November 1828; 7 August 1829, quoted in Pattison, *Statement of the Facts*, pp.5–8; 11 September 1829; 5 December 1829; 14 February 1830; undated, c. 16 February 1830; 19 June 1830; undated, July 1830; 5 September 1830; 5, 7, 9, 19 and 23 March 1831; 7 May 1831; 25 June 1831; 29 July 1831; 23 November 1831; 16 December 1831; 13 April 1832.

————to the council of the University of New York, 28 August 1831. Gallatin Collection 1831, no.28, New York Historical Society.

————to Father John DuBois, 11 October 1823. Archives, Daughters of Charity, Saint Joseph's Provincial House, Emmitsburg, Maryland.

————to Robley Dunglison, 24 December 1835. MS 506, Archives of the New York Academy of Medicine.

————to Albert Gallatin, 27 April 1827; 11 May 1827; 28 August 1831. The New York Historical Society.

————to Robert Gilmour, 8 March 1822. Dreer Collection, Physicians and Surgeons, Historical Society of Pennsylvania, Philadelphia.

————to the Rev. Dr J. M. Mathews, 27 February 1841. Simon Gratz Collection, case 7, box 32, Historical Society of Pennsylvania, Philadelphia.

————to Jonathan Meredith, 1826 undated, under the address 'Off the Capes of Delaware'. Society Collection, Granville Sharp Pattison, Historical Society of Pennsylvania, Philadelphia.

————to Robert Scott Moncrieff, 2 February 1814. Scottish Record Office, Edinburgh, AD 14/14/29.

————to Valentine Mott, 10 June 1834. National Library of Medicine, Bethesda, Maryland.

————to the Trustees of the University of Pennsylvania, 4 October 1819. Archives of the University of Pennsylvania.

Pattison, John of Kelvingrove, Trust Disposition and Deed of Settlement, Scottish Record Office, Edinburgh, Reference 4831 (1807).

Pattison, John to the editor of the *Glasgow Citizen*, 5 January 1848, quoted in the *Burns Chronicle and Club Directory*, no.4 (January 1895): 38–40.

————to his mother, Paris, 7 March 1803. In the author's possession.

————to Granville Sharp Pattison, 17 November 1818. Quoted in Pattison, *A Refutation*, pp.6–7.

Peart, J. H. to the council of the University of London, 23 February 1831.

Petition for Procurator Fiscal and Minutes and Warrants against G. S. Pattison and others, 14 and 15 December 1813. Scottish Record Office, Edinburgh, AD 14/14/29.

Philadelphia City Directories, 1820–22.

Porter, Edmund to Thomas Miner, 25 October 1825. Quoted in the *Pennsylvania Magazine of History and Biography* 16 (1892): 249.

Precognition against Messrs. Pattison and others, 202pp., 14 December 1813. Scottish Record Office, Edinburgh, AD 14/14/29.

Precognition against G. S. Pattison, etc. at instance of Mr. McAllaster and Procurator Fiscal, 46pp., 18 December 1813. Scottish Record Office, Edinburgh, AD 14/14/29.

Rainy, Harry to William Mackenzie, 12 February 1817; undated, postmarked 9 March 1819; undated, postmarked 31 July 1819; 7 August 1819. Archives of the Royal College of Physicians and Surgeons of Glasgow.

Rayner, John to the Right Honourable Lord King, 26 March 1831.

Ramshorn Burying Ground Lair Registers and Transfers, 1817–72. Mitchell Library, Glasgow, B831391.

———, 1818–46. Mitchell Library, Glasgow, B831392.

Ramshorn and Blackfriars (St. David's etc.) Burials, 1855–72. Mitchell Library, Glasgow, B831393.

Ramshorn Churchyard, Plan of, *c.* 1840. Glasgow City Archives, DT-C 13/606.

'Regulations of the Baltimore Infirmary (Patients) and Duties of Students.' Archives, Daughters of Charity, Saint Joseph's Provincial House, Emmitsburg, Maryland.

Report from the Select Committee on Anatomy, House of Commons, London, 22 July 1828.

Report of Professors Conolly, Davis and Pattison, 10 February 1830.

Report of Professors Conolly, Thomson, Turner and Davis, 9 December 1829.

Report to the Committee of the Council of the University of London, for the investigation of the complaints of the students relative to Professor Pattison, Eisdell's handwriting, signed by Henry Thomas, 16 April 1831.

Russel, Andrew, Declaration of, 29 December 1813. Scottish Record Office, Edinburgh, Small Papers, JC 26/368.

Scots Magazine, vol.4 (March 1742), vol.76 (July 1814).

Scott, General Winfield to Thomas Cadwalader, 15 April 1823. Quoted in Wainwright, 'Affair', p.344.

Shanks, Robert A., 'The Royal Medico-Chirurgical Society of Glasgow', unpublished lecture.

Somervell, William A., 'Notes upon Professor Pattison's Lectures', 1820. Archives and Library of the Medical and Chirurgical Faculty of the State of Maryland, Baltimore.

'Special Committee Report', 2 July 1830.

Spectator, vol.3, 13 February 1830, p.100; 20 February 1830, p.116; 27 February 1830, pp.133–34.

Statement by the Council of the University of London, explanatory of the Nature and Objects of the Institution (London: Longman, Rees, Orme, Brown and Green; Murray, 1827).

Student (unidentified) at New York University, 'Album of notes of a course of lectures on the practice of medicine and surgery, given by Drs. Pattison, Revere, Mott and Draper (1843–44)'. Archives of the New York Academy of Medicine.

Student (unidentified) at New York University, 'Lectures by Professor Mott, Paine, Revere, Pattison and Bedford, during the session 1843/44 of University of New York'. Archives of the New York University Medical Center.

Sun, 21, 22 and 27 April 1830.

The Times, 14 and 19 July 1827, 5 July 1830, 4 July 1831.

Timour, John A., Librarian, Thomas Jefferson University, Philadelphia, private communications to the author, 28 August and 6 September 1979.

Turner, Edward and Anthony Todd Thomson, *Two Letters to the proprietors of the University of London, in reply to some remarks in Mr. Pattison's Statement* (London: Mallett, 28 and 31 August, 1831).

University of London, 'Applicants for Situations'.

University of London Council, Minutes, vol.1 (1825–29), vol.2 (1829–35). *Volume 1*. 12 July 1827, pp.97–98; 13 December 1827, p.129; 9 February 1828, p.145; 22 November 1828, p.246; 29 August 1829, p.349; 7 September 1829, p.351; 5 December 1829, pp.372–74; 12 December 1829, p.376. *Volume 2*. 13 May 1830, p.76; 22 May 1830, p.82; 5 July 1830, p.104; 7 September 1830, p.138; 12 February 1831, p.203; 19 February 1831, p.205; 19 February 1831, p.208; 26 February 1831, p.213; 12 March 1831, p.224; 16 March 1831, p.232; 16 March 1831, p.235; 26 March 1831, p.248; 7 April 1831, p.252; 27 April 1831, p.265; 30 April 1831, p.270; 30 April 1831, p.271; 16 May 1831, p.276; 18 May 1831, p.278; 6 June 1831, p.283; 23 July 1831, p.308; 10 December 1831, p.330.

University of London, *Distribution of the Prizes and Certificates of Honours in the Medical Classes*, Saturday, 15 May 1830.

University of Maryland, Minutes of the Executive Committee, 1826–39, 24 July 1826, 9 June 1827, 7 March 1837. Special Collections, Health Sciences Library, University of Maryland, Baltimore.

————, Minutes of Faculty of Physic, 1813–37, 12 and 16 November 1824. Special Collections, Health Sciences Library, University of Maryland, Baltimore.

University of New York, Annual Announcement of Lectures, Medical Department, session MDCCCXLI–XLII; session MDCCCXLII–XLIII.

University of New York, Minutes of the Medical Faculty, 12 November 1851.

University of Pennsylvania, Minutes of the Trustees, 6 April, 6 July, 7 September, 5 October, 12 October, 1819. The minutes are located in the office of the secretary of the trustees.

University of Pennsylvania, 'Report of Committee of the Trustees', 8 October 1819. Archives of the University of Pennsylvania.

Ure Divorce, Documents relating to: Commissary Court Records, CC (Commissary Court) 8/5/37, CC 8/6/117 and CC 9/5/136. Scottish Record Office, Edinburgh. The summons, CC 8/6/117. Condescendence for the pursuer, 8 January 1819, CC 8/6/117. Pursuer's oath of calumny, CC 8/6/117. Reclaiming petition of defender, signed on 5 February 1819, called on 12 February 1819, CC 8/6/117. Scroll Decreet of Divorce, 5 February 1819 and 26 March 1819, CC 8/5/37. Catherine Monteath or Ure to Granville Sharp Pattison, 14 August 1818, CC 8/6/117. Andrew Ure to Catherine Monteath or Ure, 12 October 1818, CC 9/5/136. Catherine Monteath or Ure to Mary Park, 23 October 1818, CC 8/6/117. Catherine Monteath or Ure to Messrs. Scott and Rymer, Solicitors, 9 March 1819, CC 8/6/117. Catherine Monteath or Ure, Affidavit before Robert McLachlan, J.P., 17 March 1819, CC 8/6/117. Suit of Granville Sharp Pattison against Andrew Ure, 4 October 1819, CC 9/5/136.

Walker, David M., 'Note on Punishment for Violation of Sepulchres', private PC communication to the author, 23 July 1977.

Watson, George, handwritten notes on the case of Jean Gowdie, 13 December 1816. Archives of the Glasgow Royal Infirmary.

Whilldin, John Galloway, 'On the Nature and Treatment of that State of Disorder generally called Dropsy'. Van Pelt Library, University of Pennsylvania.

Wisconsin Tribune, 15 December 1848.

Books, Articles, and Pamphlets

Adams, Norman, *Dead and Buried? The Horrible History of Bodysnatching* (Aberdeen: Impulse Books, 1972).

Addison, W. Innes, *The Matriculation Albums of the University of Glasgow, 1728–1858* (Glasgow: Maclehose, 1913).

Agnew, L. R. C., 'Scottish Medical Education', in *The History of Medical Education*, ed. C. D. O'Malley, UCLA Forum in Medical Sciences no.12 (Los Angeles: University of California Press, 1970), pp.251–61.

American Medical Recorder 2 (1819): 614; 4 (1821): 382, 609.

Asmodeus; or Strictures on the Glasgow Democrats (Glasgow: Niven, 1793).

Ball, J. M., *The Sack-'em-up Men* (London and Edinburgh: Oliver and Boyd, 1928).

Ballard, Margaret Byrnside, *A University is Born* (Old Hundred Union, West Virginia, 1965).

Bauer, Edward Louis, *Doctors Made in America* (Philadelphia and Montreal: Lippincott, 1963).

Bayle, A. L. J. and H. Hollard, *A Manual of General Anatomy, or, a Concise Description of the Primitive Tissues and Systems which Compose the Organs in Man*, translated by Henry Storer (London: Wilson, 1829).

Bell, Charles, *Letters of Sir Charles Bell* (London: Murray, 1870).

Bellot, H. Hale, *University College, London, 1826–1926* (London: University of London Press, 1929).

Bennett, James R., *Lecture introductory to the course of general anatomy* (London: Taylor, 6 October 1830).

————, *London Medical Gazette* 7 (1831): 246.

Bishop, W. J., *The Early History of Surgery* (London: Scientific Book Guild, 1961).

Blair, George, *Biographic and Descriptive Sketches of Glasgow Necropolis* (Glasgow: Ogle, Murray, 1857).

Brooks, Thomas, *The Crown and Glory of Christianity* (1662).

Brown, John, *Horae Subsecivae* (London: Black, 1897).

Buchanan, Moses, S., *History of the Glasgow Royal Infirmary* (Glasgow: Lumsden, Brash and Robertson; Edinburgh: Black; London: Longman, 1832).

Burns, Allan, *Observations on some of the Most Frequent and Important Diseases of the Heart; on Aneurysm of the Thoracic Aorta; on Preternatural Pulsation in the Epigastric Region; and on the Unusual Origin and Distribution of some of the Large Arteries of the Human Body*, Illustrated by Cases (Edinburgh: Muirhead, 1809).

————, *Observations on the Surgical Anatomy of the Head and Neck, Illustrated by Cases and Engravings*, 2nd ed., with a life of the author and additional cases and observations by Granville Sharp Pattison (Glasgow: Ward-law and Cunninghame; London: Longman, Hurst, Rees, Orme, Brown and Green, 1824). (The first American edition was identical. Baltimore: Lucas, jr., Coale, and Cushing and Jewett; Philadelphia: Carey and Lea, 1823.)

Busey, Samuel C., *Personal Reminiscences and Recollections* (Washington, D.C., 1895).

Caldwell, Charles, *Autobiography of Charles Caldwell, M.D.*, with a preface, notes and appendix by Harriot W. Warner (Philadelphia: Lippincott, Grambo, 1855). Reprinted, with a new introduction by Lloyd G. Stevenson (New York: Da Capo, 1968).

Callcott, George H., *A History of the University of Maryland* (Baltimore: Maryland Historical Society, 1966).

Cameron, Sir Charles, *History of the Royal College of Surgeons in Ireland* (Dublin: Fannin, 1886).

Campbell, J. Menzies, *Dentistry Then and Now* (Glasgow: privately printed, 1981).

————, 'James Scott, c. 1770–1828, First Resident Dentist in Glasgow', *The Dental Magazine and Oral Topics* (December 1949): 2–16.

Cartwright, Frederick F., *The Development of Modern Surgery* (London: Barker, 1967).

————, *A Social History of Medicine* (London and New York: Longman, 1977).

Chamberlain, Joshua L., ed., *Universities and Their Sons: New York University, 1830–1900* (Boston: Herndon, 1901).

Chambers, Robert, ed., *The Life and Works of Robert Burns*, 4 vols. (New York: Harper, undated), vol.4.

Chapman,Nathaniel, *Case of Divorce of Andrew Ure, M. D. v. Catharine Ure* (Philadelphia: Fry, 1 September 1821).

——, *Correspondence between Mr. Granville Sharpe Pattison and Dr. N. Chapman*, 5 November 1820, 2nd ed. (Philadelphia: August 1821).

Chase, Heber, *The Medical Student's Guide* (Philadelphia: Auner, 1842).

Cleland, James, *The History of the High School of Glasgow, containing the historical account of the Grammar School* (Glasgow: Bryce, Lumsden, 1878).

Clubbe, John, ed., *Selected Poems of Thomas Hood* (Cambridge, Mass.: Harvard University Press, 1970).

Cockburn, Lord Henry, *Life of Lord Jeffrey, with a Selection from his Correspondence*, 2 vols. (Philadelphia: Lippincott, 1856).

Colles, Abraham, *A Treatise on Surgical Anatomy* (Dublin: Gilbert and Hodges, 1811).

Comrie, J. D., *History of Scottish Medicine*, 2 vols. (London: Ballière, Tindall and Cox, for the Wellcome Historical Medical Museum, 1932).

Cooper, Bransby Blake, *The Life of Sir Astley Cooper, Bart.*, 2 vols. (London: Parker, 1843).

Cope, Zachary, *The Royal College of Surgeons of England: A History* (London: Blond, 1959).

Copeman, W. S. C., 'Andrew Ure, M.D., F. R. S. (1778–1857)', *Proceedings of the Royal Society of Medicine* 44, no.8 (1951): 655–62.

Cordell, Eugene Fauntleroy, *Historical Sketch of the University of Maryland School of Medicine (1807–1890)* (Baltimore: Friedenwald, 1891).

——, *University of Maryland, 1807–1907*, 2 vols. (New York and Chicago: Lewis, 1907).

Corner, George W., *Two Centuries of Medicine: A History of the School of Medicine, University of Pennsylvania* (Philadelphia and Montreal: Lippincott, 1965).

Coutts, James, *A History of the University of Glasgow: from its foundation in 1451 to 1909* (Glasgow: Maclehose, 1909).

Cowen, David L., 'George Bushe and the "Rutgers Medical Faculty"', *Journal of the Rutgers University Library* 30 (1966): 1–7.

Daiches, David, *Glasgow* (London: Deutsch, 1977).

Dale, William, *The State of the Medical Profession in Great Britain and Ireland* (Dublin: Fannin; London: Longmans, 1869).

Daughter of Charity, A, *Mother Rose White* (Emmitsburg, Maryland: St Joseph's, 1936).

'Death of Prof. Granville Sharpe Pattison', *New York Journal of Medicine* 8 (1852): 143.

de Morgan, Sophia Elizabeth, *Memoir of Augustus de Morgan* (London: Longmans, Green, 1882).

Dictionary of National Biography, ed. S. Lee (London: Smith and Elder, 1909).

Duncan, Alexander, *Memorials of the Faculty of Physicians and Surgeons of Glasgow, 1599–1850* (Glasgow: Maclehose, 1896).

Dunglison, Robley, 'The Autobiographical Ana of Robley Dunglison', ed. Samuel X. Radbill, *Transactions of the American Philosophical Society*, n.s., 53 part 8 (1963).

————, *The Medical Student; or, Aids to the Study of Medicine* (Philadelphia: Lea and Blanchard, 1844).

The Emmet, a selection of original essays, tales, anecdotes, bon mots, choice sayings, etc., 2 vols. (Glasgow: Purvis and Aitken, 1824).

Evidence, oral and documentary, taken and received by the Commissioners appointed by His Majesty George IV, 23 July 1826, and reappointed by His Majesty William IV, 12 October 1830, for visiting the Universities of Scotland. Vol.2, University of Glasgow. Presented to both Houses of Parliament by Command of His Majesty. (London: Clowes, 1837).

Eyre-Todd, George, *History of Glasgow* (Glasgow: Jackson, Wylie, 1934).

Farrar, W. V., 'Andrew Ure, F.R.S. and the philosophy of manufactures', *Notes and Records of the Royal Society of London* 27, no.2 (1973): 299–324.

Ferguson, J. De L., *Letters of Robert Burns*, 2 vols. (Oxford: Clarendon Press, 1931).

Fergusson, W., 'The Operation of Lithotomy', *New York Lancet* 3 (1843): 1–5.

Figge, Frank H. J., 'Granville Sharp Pattison, the Dueling Anatomist', *Bulletin of the School of Medicine, University of Maryland* 23, no.2 (1938): 81–92.

Flexner, Abraham, *Medical Education in the United States and Canada: a report to the Carnegie Foundation* (New York: Carnegie Foundation, 1910).

Fyfe, Andrew, *The Anatomy of the Human Body, illustrated by 158 plates, taken partly from the most celebrated authors, partly from Nature* (Edinburgh: Black; London: Longmans, Green, 1830).

Garnett, Thomas, *Observations on a Tour through the Highlands and Part of the Western Isles of Scotland*, 2 vols. (London: Cadell and Davies, 1800).

Gathorne-Hardy, Jonathan, *The Rise and Fall of the British Nanny* (London, Sydney, Auckland and Toronto: Hodder and Stoughton, 1972).

Gayley, James F., *A History of the Jefferson Medical College of Philadelphia* (Philadelphia: Wilson, 1858).

Gibson, William, *Strictures on 'Mr. Pattison's Reply to Certain Oral and Written Criticisms'* (Philadelphia: Maxwell, 1820).

————(writing under the pseudonym 'W'), 'Observations on Mr. Pattison's Paper on the Operation of Lithotomy, etc.', *American Medical Recorder* 3 (1820): 252–59.

Goodall, A. L., 'Granville Sharp Pattison: the Argumentative Anatomist', *Proceedings of the Scottish Society of the History of Medicine* 1959: 20–23.

Goodman, Lee Dana, 'Grave Robbing, a once prevalent and profitable undertaking', *Pharos* (Alpha Omega Alpha Honor Medical Society) 39, no.3 (1976): 86–87.

Gordon, James, ed., *Glasghu Facies* (Glasgow: John Tweed, 1872).

Gould, George M., *The Jefferson Medical College of Philadelphia, 1826–1904: A History* (New York and Chicago: Lewis, 1904).

Graham, Harvey, *Surgeons All* (London: Rich and Cowan, 1939). (American edition: *The Story of Surgery* [New York: Doubleday, 1939]).

Graham, Henry Grey, *The Social Life of Scotland in the Eighteenth Century* (London: Black, 1909).

Gross, Samuel D., *Autobiography of Samuel D. Gross, M.D., with Sketches of his Contemporaries*, edited by his sons, 2 vols. (Philadelphia: Barrie, 1887).

———, 'A Century of American Medicine. 1776–1876. II. Surgery', *American Journal of the Medical Sciences*, n.s., 71 (1876): 437.

Gunson, E. Sherwood, *The Story of the Ramshorn Churchyard* (privately printed, c. 1910). Mitchell Library, Glasgow; G285.2, B446190.

Hamilton, David, *The Healers: a history of medicine in Scotland* (Edinburgh: Canongate, 1981).

Harding, Chester, *A Sketch of Chester Harding, Artist, drawn by his own hand*, ed. Margaret E. White, annotations by W. P. G. Harding (New York: Kennedy Galleries, Da Capo Press, 1970).

Harte, Negley and John North, *The World of University College London 1828–1978* (London: University College; Portsmouth: Eyre and Spottiswoode, 1978).

Harvey, Warren, *Dental Identification and Forensic Odontology* (London: Kimpton, 1976).

Heaton, Claude Edwin and Allan Eliot Dumont, *The First One Hundred and Twenty-Five Years of the New York University School of Medicine* (New York: New York University and New York School of Medicine, 1966).

Henry, F. P., ed., *Standard History of the Medical Profession of Philadelphia* (Chicago: Goodspeed, 1897).

Herrick, James B., 'Allan Burns: 1781–1813. Anatomist, Surgeon and Cardiologist', *Bulletin of the Society of Medical History of Chicago* 4 (1935): 457–83.

Hood, Thomas, *The Choice Works of Thomas Hood in Prose and Verse* (London: Chatto and Windus, 1897).

Horner, Leonard, *Letter to the Council of the University of London* (London: Moyes, 1 June 1830).

———, *Memoir of Leonard Horner*, 2 vols., ed. Katharine M. Lyell, his daughter (London: Women's Printing Society, 1890).

Houston, Robert, 'An account of a dropsy in the left ovary of a woman, aged 58, cured by a large incision made in the side of the abdomen', *Philosophical Transactions of the Royal Society* 33 (1724): 8–15.

Hume, David, *Commentaries on the Law of Scotland, Respecting the Description and Punishment of Crimes* 1st ed., 2 vols. (Edinburgh: Bell and Bradfute and Balfour, 1797); 2nd ed., 1819; 4th (last) ed., 1844.

Hunter, Richard H., *A Short History of Anatomy*, 2nd ed. (London: Bale, Danielsson, 1931).

Jameson, Horatio, G., 'Additional observations on the case of James Underwood', *American Medical Recorder* 6 (1823): 773–74.

——, 'Case of Tumour of the Superior Jaw', *American Medical Recorder* 4 (1821): 222–30.

——, Certificates relating to the health of James Underwood, *American Medical Recorder* 16 (1829): 229–32.

——, ('H. J.'), 'An examination of a case of Aneurism, communicated by Professor Pattison', *American Medical Recorder* 5 (1822): 478–88.

——, 'Observations on the parts concerned in Lithotomy, which are intended to prove that Mr. Pattison's ideas of a Prostate Fascia are erroneous', *American Medical Recorder* 5 (1820): 325–31.

Kaufman, Martin, *American Medical Education: the formative years 1765–1910* (Westport, Connecticut and London: Greenwood Press, 1976).

Kennedy, James W., *The Unknown Worshipper* (New York: Morehouse-Barlow, 1964).

Lancet, 1826–27, xii, 18. 1827–28, ii, 819. 1828–29, ii, 339. 1829–30, i, 6–8, 897–98, 931. 1829–30, ii, 623–24, 740–45, 799–800, 847–48, 974–75. 1830–31, i, 186, 286–87, 749–53, 815–18. 1830–31, ii, 15, 123, 149, 192, 209–11, 311, 470–73, 689–95; 721–27, 733–34, 735–36, 744–50, 753–57, 763–64, 785–90, 793–95, 825–29. 1831–32, i, 82–87, 185–89, 209–15, 812.

Lascelles, E. C. P., *Granville Sharp and the Freedom of Slaves in England* (Oxford University Press; London: Milford, 1928).

Lindsay, Maurice, *The Burns Encyclopedia* (London: Hutchinson, 1959).

Lipton, Leah, 'Chester Harding in Great Britain', *Antiques* 125 (1984): 1382–90.

Lockhart, John Gibson, *Peter's Letters to his Kinsfolk*, 3 vols. (Edinburgh: Blackwood; London: Cadell, Davies; Glasgow: Smith, 1819).

Lomas, Arthur J., 'As it was in the Beginning: A History of the University Hospital', *Bulletin of the School of Medicine, University of Maryland* 23, no.4 (1939): 182–209.

London Medical Gazette 7 (1831): 822–24; 8 (1831): 692–93 (leading article).

Lyle, Thomas, *Ancient Ballads and Songs* (London: Relfe; Dublin: Westley and Tyrrel; Edinburgh: Constable; Glasgow: Lumsden, 1827).

Mackenzie, Peter, *Old Reminiscences of Glasgow and the West of Scotland*, 3 vols. (Glasgow: Forrester, 1890).

McLehose, W. C., ed., *The Correspondence between Burns and Clarinda* (New York: Bixby, 1843).

McLellan, Duncan, *Glasgow Public Parks* (Glasgow: Smith, 1894).

Maitland, Frederick Lewis, *The Surrender of Napoleon*, ed. W. K. Dickson (Edinburgh and London: Blackwood, 1904).

Medico-Chirurgical Review and Journal of Practical Medicine 15 (October 1831): 330–36; 16 (January 1832): iv.

Meikle, Henry W., 'Two Glasgow Merchants in the French Revolution', *Scottish Historical Review* 8 (1911): 149–58.

Merrington, W. R., *University College Hospital and its Medical School: a History* (London: Heinemann, 1976).

Miles, W. D., 'Washington's First Medical Journal: Duff Green's Register and Library of Medical and Chirurgical Science, 1833–36', *Records of the Columbia Historical Society* (1969–70): 114–25.

Miller, William Snow, 'Granville Sharp Pattison', *Johns Hopkins Hospital Bulletin* 30 (1919): 98–104.

Moncrieff, Frederick and William Moncreiffe, *The Moncrieffs and the Moncreiffes*, 2 vols. (Edinburgh: privately printed, 1929).

Moore, James, *A Method of Preventing or Diminishing Pain in Several Operations of Surgery* (London: Cadell, 1784).

Moore, Norman and Stephen Paget, *History of the Royal Medical and Chirurgical Society of London Centenary, 1805–1905* (Aberdeen: Aberdeen University Press, 1905).

Muir, James, *John Anderson, Pioneer of Technical Education, and the College He Founded* (Glasgow: Smith, 1950).

Murray, David, *Memories of the Old College of Glasgow: Some Chapters in the History of the University* (Glasgow: Jackson, Wylie, 1927).

Murray, Thomas, *Autobiographical Notes, also Reminiscences of a Journey to London in 1840*, ed. John A. Fairley (Dumfries: Standard Office, 1911).

Newman, Charles, *The Evolution of Medical Education in the Nineteenth Century* (London, New York and Toronto: Oxford University Press, 1957).

New York Lancet 1 (1842): 11–13, 27, 64 (opposite), 218, 219, 235, 251, 268, 284, 300, 304, 315; 2 (1842): 25–28, 42, 185–87, 283.

Nichol, *Glasgow Illustrated in Twenty One Views* (Montrose: J. & D. Nichol, 1841).

Northern Looking Glass (Glasgow: Watson, 1825).

Norwood, William Frederick, 'Medical Education in the United States before 1900', in *The History of Medical Education*, ed. C. D. O'Malley, UCLA Forum in Medical Sciences no. 12 (Los Angeles: University of California Press, 1970).

————, *Medical Education in the United States before the Civil War* (Philadelphia: University of Pennsylvania Press, 1944).

Odell, George, C. D., *Annals of the New York Stage* (New York: Columbia University Press, 1928, 1931), vols. 4 (1834–43), 5 (1843–50) and 6 (1850–57).

The Old Country Houses of the Old Glasgow Gentry, 2nd ed. (Glasgow: Maclehose, 1878).

Osler, Sir William, *An Alabama Student and Other Biographical Essays* (New York: Oxford University Press; London: Frowde, 1909).

'The Park District: a Mid-Victorian Achievement', *Glasgow Chamber of Commerce Journal* 50, no. 7 (1965): 439–41.

Pattison, Frederick Hope, *Personal Recollections of the Waterloo Campaign, in a series of letters to his grand-children* (Glasgow: printed for private circulation, 1873).

Pattison, F. L. M., 'The Clydesdale Experiments: an early attempt at resuscitation', *Scottish Medical Journal*, 31 (1986): 50–52.

————, 'The Pattison–Miller Quarrel', *Scottish Medical Journal* 25 (1980): 234–40.

————, 'Uncle Granville', *University of Western Ontario Medical Journal* 37 (1966): 24–27.

Pattison, Granville Sharp: a full list of works by Pattison appears as Appendix 1.

Pike, E. Royston, *Human Documents of the Industrial Revolution in Britain* (London: Allen and Unwin, 1966).

Potter, Nathaniel, *Some Account of the Rise and Progress of the University of Maryland* (Baltimore: Robinson, 1838).

Quinan, John R., *Medical Accounts of Baltimore from 1608 to 1880* (Baltimore: Friedenwald, 1884).

Ramsey, Ted, *Don't Walk Down College Street* (Glasgow: Ramshorn, 1985).

Raper, H. R., *Man Against Pain: the Epic of Anesthesia* (New York: Prentice Hall, 1945).

'Recent Improvements in Medical Education', *Quarterly Journal of Education* 4 (1832): 1–21.

'Remarques accompagnées d'observations sur les plaies d'abdomen, par M. Grandville Sharp Pattison', *Journal Général de Médecine* 66 (1819): 388–96.

Report of the Committee Appointed by the Quarterly Meeting of the Town's Hospital, 19th November 1815 (Glasgow: Jack and Gallie, 1816).

Richman, Irwin, *The Brightest Ornament: A Biography of Nathaniel Chapman, M.D.* (Bellafonte, Pennsylvania: Pennsylvania Heritage, 1967).

Roughead, William, ed., *Burke and Hare* (Edinburgh and London: Hodge, 1921).

Satchwell, S. S., *Annual Address before the Alumni Association of the University of the City of New York, Medical Department, 4 March 1873*.

Scott, Hew, *Fasti Ecclesiae Scotiani* (Edinburgh: Oliver and Boyd, 1920).

'Senex' (Robert Reid), *Glasgow Past and Present*, 3 vols. (Glasgow: Robertson, 1884), vol.2, pp.73–75.

Seton, George, *The House of Moncrieff* (Edinburgh: privately printed, 1890).

Sexton, A. Humboldt, *The First Technical College* (London: Chapman and Hall, 1894).

Shanks, Robert A., 'Glasgow—One Hundred and Fifty Years Ago', *Scottish Medical Journal* 9 (1964): 377–83.

————, 'Granville Pattison and the Uses of History', *Scottish Medical Journal* 11 (1966): 267–76.

Sharp, Granville, *Extract from a Representation of the Injustice and Dangerous Tendency of Tolerating Slavery, or Admitting the Least Claim of Private Property in the Persons of Men in England* (London, 1769; Philadelphia: Crueshank, 1771).

Shryock, Richard Harrison, *The Development of Modern Medicine* (Philadelphia: University of Pennsylvania Press; London: Milford, Oxford University Press, 1936).

Sims, J. Marion, *The Story of My Life*, edited by his son, H. Marion Sims, M.D. (New York: Appleton, 1884).

Smith, D. Nichol, 'Thomson and Burns', *Some Observations on Eighteenth-Century Poetry* (Toronto: University of Toronto Press, 1937), reprinted in *Eighteenth-Century English Literature: Modern Essays in Criticism*, ed. James L. Clifford (New York: Oxford University Press, 1959), p.183.

'Smith, Leonard', *Northern Sketches or Characters of Gxxxxxx* (London: Dick, c. 1811).

Smout, T. C., *A History of the Scottish People, 1560–1830* (Glasgow: Collins, Fontana, 1972).

Solly, Henry, *These Eighty Years, or The Story of an Unfinished Love* (London: Simpkin, Marshall; Croydon: Hayward, 1893).

Strang, John, *Glasgow and Its Clubs* (Glasgow: Smith, 1864).

Thomas, Maurice Walton, *The Early Factory Legislation* (London: Thames Bank Publishing, 1948; Westport, Connecticut: Greenwood Press, 1970).

Thomson, A. M. Wright, *The Life and Times of Dr William Mackenzie, Founder of the Glasgow Eye Infirmary*, privately printed (Glasgow: Maclehose, 1973).

Thomson, Alexander, 'A Sop for Cerberus!' *London Medical and Surgical Journal* 5 (1830): 437–56.

————*To Lord Brougham, and through him to the proprietors of the London University, reasons why the expulsion of Alexander Thomson, M.B. from the London University, should be reconsidered, and answers to certain charges, against the character and conduct of the same gentleman, made in a recent 'statement' by Mr. Pattison, a discarded, if not disgraced professor of the London University, with testimonials of the good character, industry and progress of the said Alexander Thomson, for twenty years back* (Paris: Chez Auguste Mie, undated but after 8 October 1831).

Thomson, James, 'The Pattisons', *Burns Chronicle and Club Directory*, no.33 (January 1924): 26–31.

Villermé, L.-R., 'Rapport fait à la Société Médicale d'Émulation sur un mémoire ayant pour titre: Observations on abdominal wounds with cases, etc., par M. Granville Sharp Pattison', *Journal Universel des Sciences Médicales* 13 (1819): 241–52.

Wainwright, Nicholas B., 'Affair with Professor Pattison', *Pennsylvania Magazine of History and Biography* 64, no.3 (1940): 331–44.

Warren, J. Collins, *The Influence of Anaesthesia on the Surgery of the Nineteenth Century* (Boston: privately printed, 1906).

Whittet, T. D., 'London's University Dispensary: Torchbearer for University College Hospital, London', *The Chemist and Druggist* (1962): 217–19.

Wilson, John J., *The Annals of Penicuik* (Edinburgh: Constable, 1891).
Wordsworth, Dorothy, *Journals of Dorothy Wordsworth*, ed. E. de Selincourt, 2 vols. (London: Macmillan, 1959).

Index

Page numbers in *italics* refer to illustrations

Eisdell, Nathaniel, 156, 157, 158, 160, 163, 174–5, 176, 178, 182
electrification of corpses, 15
England
 law on grave-robbery, 242
 medical education
  early 19th century, 139–44
Emmet, The, 29
Erasistratus, 26
examinations
 at London University, 148
 at Maryland University, 119
 physical, of patients, 141
execution by hanging, 26–7
exhumations, illegal, 27–31, 115
 of Mrs McAllaster, 31–7
eye surgery, 207

facial surgery, 208
Faculty of Physicians and Surgeons of Glasgow, 13, 24–5, 73–4, 218
Faraday, Michael (1791–1867), 59
Fearon, Herbert, 165
fees, medical
 collection of, 201
 in the United States, 200
France
 and medical education, 143
 Pattison's visits to, 75
Franklin, Benjamin, 106
Frazer, Jean, 46, 47
Freer, Robert (1745–1827), 18
French anatomy, 179
Fyfe, Andrew, 162

Gallatin, Albert (1761–1849), 135–6, 198
Garnett, Thomas (1766–1802), 75
Geddings, E., 187
Gibson, John, 45
Gibson, William (1788–1868), 92, 94, 102, 107
 Pattison's quarrel with, 100–5

Gillies, Lord, 43
Glasgow
 Anderson's Institution, 73–6, 74, 84, 85–6, 245
 grave-robbery in, 28–9
 hospitals, 61
 Pattison's early days in, 1–25
 Royal Infirmary, 13, 61–4
 social conditions, 2–3
 University of, 12–20
Glasgow Grammar School, 9–12, 11, 220, 239–40
Glasgow Herald, 42, 244
Glagow Medical Society, 58–61, 59, 245
Gowdie, Jean, 66–8, 69–71, 73
Graham, Robert (1786–1845), 58, 65, 68, 72
Grant, Robert Edmond (1793–1874), 173
grave-robbery (body-snatching), 21, 27–31
 law regarding, 54–5, 115, 242
Greek, at Glasgow Grammar School, 10
Green, Jacob (1790–1841), 184
Gross, Samuel, 188, 205
Gunn, Janet, 32, 47

Hamilton, Bertha, 54
hanging, execution by, 26–7
Harding, Chester (1792–1866), 110, 231, 232
Hare, Robert (1781–1858), 106
Hare, William (c.1792–1869), 28, 244
Hegar, Alfred (1830–1914), 142
Hermand, Lord, 43
Herophilus, 26
Heurteloup, Charles (1793–1864), 152–3
History of the University of Maryland (Callcott), 114
Hoffman, Murray, 210

About the author

Fred Pattison, who holds the degrees of Ph.D. and Sc.D. (Cantab.) and M.D. (University of Western Ontario), is director of Student Health Services at the University of Western Ontario and director of the Sexually Transmitted Diseases Clinic in London, Ontario.

Born in Glasgow in 1923, he attended Loretto School and the University of Cambridge, and later moved to the University of Western Ontario, where he was professor and head of the chemistry department. At age forty-two, he resigned to become a medical student, graduating in 1969. After two years of solo family practice with the International Grenfell Association in northern Newfoundland, he returned to London, Ontario.

His main field of research in chemistry involved the synthesis, properties, and uses of biologically active, aliphatic fluorine compounds, much of which is described in his book *Toxic Aliphatic Fluorine Compounds* (Elsevier, 1959). Now, in medical research, he is studying clinical and pharmacological aspects of the sexually transmitted diseases.

He has lectured extensively in many countries of the world, and has made several trips to the High Arctic, living with the Inuit under extremes of cold to study the local hobbies and problems of survival. He is a keen musician and has given piano recitals over the Canadian radio networks.